**elementary
calculus
for
business,
economics,
and
social
sciences**

elementary calculus for business, economics, and social sciences

Chaney Anderson
R. C. Pierce, Jr.
University of Houston

HOUGHTON MIFFLIN COMPANY · BOSTON
Atlanta · Dallas · Geneva, Ill.
Hopewell, N.J. · Palo Alto · London

To Jo Jean and Linda

Library of Congress Catalog Card Number: 74-14361

ISBN: 0-395-18960-8

contents

preface

This text is designed for a one-semester course in calculus for students of business and the liberal arts. Our intention is to present the necessary topics of calculus in an intuitive manner appropriate for a first course in college mathematics.

The prerequisite for this course should be a knowledge of high-school algebra, but no trigonometry is required.

Some of the advantageous features of this book are as follows:

1 Each topic is illustrated with several examples and nonexamples. A non-example illustrates something that is contrary to the topic under discussion. For example, in the discussion of functions a nonexample illustrates something that is not a function.

2 Applications for each topic are given as the topic is developed.

3 The basic rules of differentiation and their applications are presented before the student is required to learn the more complex rules and their applications.

4 Integration is presented as the opposite of differentiation so that the student can rely on his knowledge of differentiation.

5 The number e and its applications are introduced early in the text. This is accomplished without the use of logarithms.

6 Logarithms are treated completely, from the definition of a logarithm to the derivatives and integrals of logarithms. Some applications of logarithms are also given.

Since the aim of this course is to develop calculus as a tool to be used in solving certain types of problems, we have tried to minimize the technical difficulties that arise in a rigorous study of calculus. In order to do this, we have restricted our attention to situations that exemplify the basic techniques with a minimum of theoretical difficulties.

The authors feel that the somewhat unusual ordering of the topics is justified. For example, the topic of exponential functions presents little difficulty to students, while the topic of logarithms is often an enigma. Hence the exponential function is separated from logarithms, even though the treatment is then not as neat.

Also, problems requiring maximums and minimums are presented in three different chapters. The first encounter, in Chapter 3, does not require the students to form their own equations; therefore they can concentrate on learning the new concepts involved in the solutions. The second time that maximums and minimums are discussed, in Chapter 4, the students are usually asked to supply their own equations, and more complex differentiation formulas are given. The third discussion, in Chapter 7, involves higher derivatives and functions of several variables.

The authors wish to thank the following professors for reviewing the manuscript and offering helpful suggestions: Melvin F. Janowitz, University of Massachusetts; Tyler Haynes, Catonsville City College; Stanley M. Lukawecki, Clemson University; E. D. Davis, State University of New York at Albany; and Phillip Gillett, University of Wisconsin—Marathon Center.

Chaney Anderson
R. C. Pierce, Jr.

table of symbols

$=$	means equal
$\doteq$	means approximately equal
$a < b$	means a is less than b
$a \leq b$	means a is less than or equal to b
$a > b$	means a is greater than b
$a \geq b$	means a is greater than or equal to b
x^a	is a power, where a is the exponent and x is the base
$y = f(x)$	means y is a function of x
(a, b)	represents an ordered pair
$\left. \begin{array}{l} dy/dx \\ f'(x) \end{array} \right\}$	means the derivative taken with respect to x
$\int$	means find the indefinite integral
$\int_a^b$	means find the definite integral from a to b
π	is 3.14159 . . .
e	is 2.71828 . . .
$x \rightarrow a$	means x approaches a
∞	means infinity

introduction

Students majoring in business and other fields have witnessed a steady influx of higher mathematics into their disciplines in recent years. Today, such mathematical concepts are vital in many practical fields; they are becoming more indispensable with each passing year.

Since many students do not pursue a mathematics curriculum in their pre-college studies, the first chapter of this book stresses certain topics of algebra and geometry that are especially useful in the study of calculus. The background material that is given in Chapter 1 should aid the student in understanding the concepts that are presented later.

Through the years many students have come to regard calculus as a dreaded, sometimes fatal, disease. The authors want to present calculus in small, painless doses, but still strong enough to help in solving certain types of problems.

The first question that comes to mind in regard to calculus is, What is calculus? But it seems to us that a better question might be, What can we do with calculus? In Chapters 3, 4, 7, and 8 the branch of calculus called *differential calculus* is discussed. Differential calculus serves in two main ways. First, we are able to take complicated functions and determine their maximum or minimum value. This is useful in solving the following problems:

1 At what level of output will I have a maximum profit?

2 At what level of output will I have a minimum average cost?

3 What dimensions should I use in order to make a container that will hold a fixed volume and yet minimize cost?

4 What should be the capacity of this facility in order to maximize revenue?

Second, we are able to find the rate of change of one quantity with respect to another at any given instant. The best example of this is the velocity of an object.

1

When the velocity is not constant, the derivative enables us to find the velocity at any instant. In economics, the derivative is important in marginal analysis.

In Chapters 5, 6, and 8 we study the branch of calculus called *integral calculus*. The definite integral enables us to find a bounded area even though the figure is not a standard geometrical figure. The definite integral is also useful for certain work in statistics, such as the normal distribution curve.

Relax and enjoy the material. Approach it from the viewpoint that you are being armed with some powerful concepts that will help you turn puzzling questions into simple answers.

chapter one

functions and graphs

1.1 Graphing

Graphing is useful in the study of calculus because the graph of an equation is a picture of the solutions of the equation. Often from the picture the student can observe some of the properties of the equation.

Basically, graphing is a system used to describe the location of points in a plane. In this book we will use the rectangular Cartesian coordinate system, which was invented by the French mathematician René Descartes during the seventeenth century. Its components appear in Figure 1.1.

FIGURE 1.1

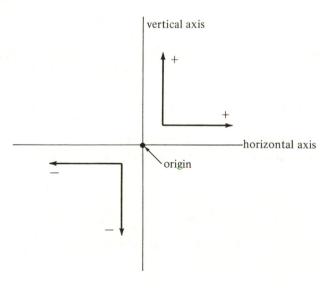

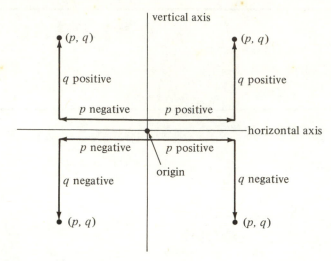

FIGURE 1.2

The rectangular Cartesian coordinate system consists of two perpendicular number lines called the *coordinate axes*. The axes are placed on a plane so that one axis is horizontal and the other vertical. All points in the plane are located with respect to the point of intersection of the axes. The point of intersection is called the *origin*. Distances measured to the right (horizontally) or upward (vertically) from the origin are designated *positive*. Distances measured to the left (horizontally) and downward (vertically) are designated *negative*.

Each point in a plane corresponds to an ordered pair of real numbers called the *coordinates* of the point. The points in Figure 1.2 have coordinates p and q, denoted (p, q). The points are located by moving p units from the origin in a horizontal direction (to the right if p is positive or to the left if p is negative) and then moving q units in the vertical direction (upward if q is positive and downward if q is negative). When we have located a point on a plane we say we have *plotted* the point.

The pair of real numbers (p, q) is called an *ordered pair* because the first number in the pair always represents a distance on the horizontal axis while the second number in the pair always represents a distance on the vertical axis. In other words, the pair (3, 2) locates a point different from the one located by (2, 3).

Example 1 Plot the following points:

$\qquad$ a. (3, 5) $\qquad$ b. (−3, 5) $\qquad$ c. (−3, −5) $\qquad$ d. (3, −5)

$\qquad$ e. (0, 2) $\qquad$ f. (2, 0) $\qquad$ g. (−2, 0) $\qquad$ h. (0, −2)

The points are plotted in Figure 1.3.

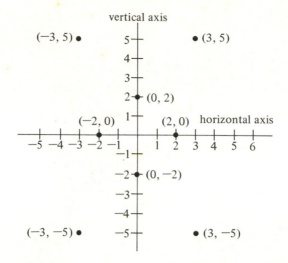

FIGURE 1.3

Applied Example 2 The costs of mailing packages weighing from 1 to 10 lb fourth class (locally) are given in Table 1.1. Rewrite the table in terms of ordered pairs. Plot the data.

In terms of ordered pairs: (1, 60), (2,60), (3, 60), (4, 65), (5, 70), (6, 70), (7, 75), (8, 75), (9, 80), and (10, 80). The points are plotted in Figure 1.4.

TABLE 1.1

Number of pounds	Cost (¢)
1	60
2	60
3	60
4	65
5	70
6	70
7	75
8	75
9	80
10	80

FIGURE 1.4

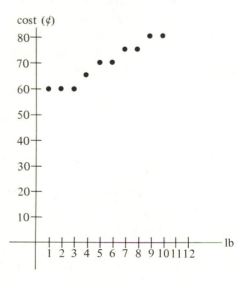

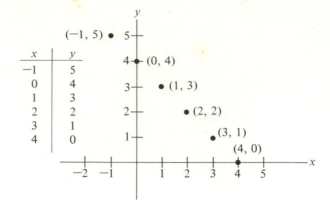

x	y
−1	5
0	4
1	3
2	2
3	1
4	0

FIGURE 1.5

Another way of obtaining ordered pairs is to use the solution pairs of equations in two variables.

Example 3 Compute some solution pairs for the equation $x + y = 4$ and plot them.

To compute some solution pairs, we assign some value to one of the variables, say x, and then compute the corresponding value of the other variable y. For each value of x we have a corresponding value for y, hence the name *solution pair*. Some solution pairs of $x + y = 4$ are given in the table accompanying Figure 1.5.

We plot the pairs with the values of x corresponding to distances from the origin on the horizontal axis and the values of y corresponding to distances from the origin on the vertical axis. In this case, we call the horizontal axis the *x axis* and the vertical axis the *y axis*.

There are infinitely many solution pairs for the equation in this example, so they cannot all be computed or plotted. However, we can draw a picture of the set of all solution pairs for the equation $x + y = 4$ by drawing a smooth line (sometimes straight and sometimes curved) through the points we have already plotted. (See Figure 1.6.) This drawing of the line through the points we have plotted is called *graphing* the equation. It is implied that all the points on the smooth line are other solutions of the equation $x + y = 4$ (unless specifically stated to the contrary). That is, choose any point on the smooth line and the coordinates of that point will be a solution pair of the equation $x + y = 4$.

The point where the graph crosses the x axis is called the *x intercept*. This occurs when $y = 0$. In Example 3 the x intercept is 4.

The point where the graph crosses the y axis is called the *y intercept*. This occurs when $x = 0$. In Example 3 the y intercept is 4.

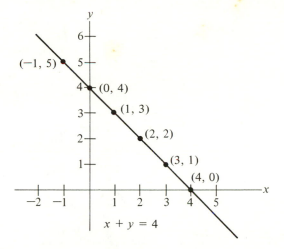

FIGURE 1.6

Example 4 Graph the equation $y = 2x + 1$, where the horizontal axis is the x axis and the vertical axis is the y axis.

To compute some representative solution pairs, we assign some values to one variable, say x, and compute the corresponding values of y.

The x intercept is $-\frac{1}{2}$. The y intercept is 1. (See Figure 1.7.) An equation that has the form (or can be put into the form) $y = mx + b$, where x, b, and m are real numbers, is graphed as a straight line.

FIGURE 1.7

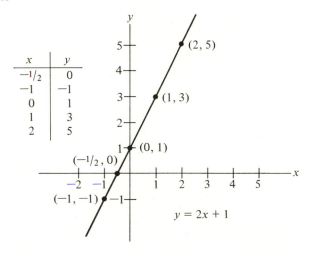

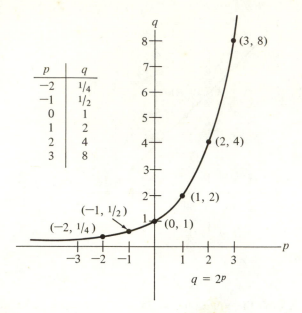

p	q
-2	1/4
-1	1/2
0	1
1	2
2	4
3	8

$q = 2^p$

FIGURE 1.8

Example 5 Graph the equation $q = 2^p$, where the horizontal axis is the p axis and the vertical axis is the q axis. Find the intercepts of the graph if there are any. (*Note:* An equation of the type $q = 2^p$ is called an *exponential equation*.)

As may be seen from Figure 1.8, the q intercept is 1, there is no p intercept. Note that the locations of the points usually indicate how the smooth curve is to be drawn between the points.

FIGURE 1.9

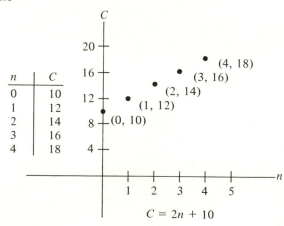

n	C
0	10
1	12
2	14
3	16
4	18

$C = 2n + 10$

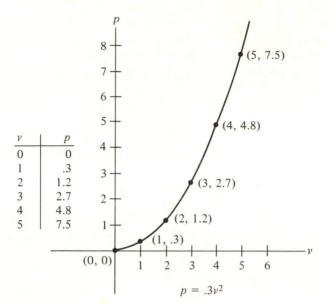

v	p
0	0
1	.3
2	1.2
3	2.7
4	4.8
5	7.5

$$p = .3v^2$$

FIGURE 1.10

Applied Example 6 Let n be the number of units of an item and C be the cost in dollars of producing that item, as given by the equation $C = 2n + 10$. Graph the equation, letting the horizontal axis be the n axis and the vertical axis be the C axis.

The graph is shown in Figure 1.9. Only nonnegative integers were used for n because a negative number of units produced is not meaningful.

Applied Example 7 The force p due to wind resistance against a moving object is related to the velocity v of the object, as given by the equation $p = .3v^2$. Graph the equation, letting the horizontal axis be the v axis and the vertical axis be the p axis.

The graph is shown in Figure 1.10.

▶ **Exercise 1.1**

Plot the points in Problems 1 through 6.

1 a $(1, 3)$ **b** $(-1, 3)$ **c** $(-1, -3)$ **d** $(1, -3)$
2 a $(0, 4)$ **b** $(4, 0)$ **c** $(-4, 0)$ **d** $(0, -4)$
3 a $(0, 0)$ **b** $(3, 3)$ **c** $(-2, -2)$ **d** $(1, 5)$
4 a $(2, 3)$ **b** $(2, 5)$ **c** $(2, -4)$ **d** $(2, 0)$
5 a $(-2, 5)$ **b** $(-2, 3)$ **c** $(-2, 0)$ **d** $(-2, \frac{1}{2})$
6 a $(2.5, 3)$ **b** $(-1.5, 3.5)$ **c** $(2\frac{1}{2}, 3\frac{1}{4})$

7 A teacher wanted to know if there was any relation between the order in which her students completed a test and the score they earned. She collected the data shown in Table 1.2. Rewrite the table in terms of ordered pairs. Plot the data, labeling the horizontal axis "order of completion" and the vertical axis "grade."

TABLE 1.2

Order of completion	Grade	Order of completion	Grade
1	21	6	87
2	90	7	77
3	52	8	62
4	98	9	42
5	82	10	58

Let the horizontal axis be the x, q, t, or n axis where appropriate. Graph the equations in Problems 8 through 15.

8 $p = 2q - 3$.

9 $c = 10n + 100$.

10 $y = 3^x$.

11 $y = 3x - 4$.

12 $y = x - 3$.

13 $R = 4t - t^2$.

14 $y = 10^x$.

15 $y = (\frac{1}{2})^x$.

16 A company knows that its total cost C of producing x items is given by $C = 5x + 20$. Graph the equation. What is the cost of producing 20 items?

17 A firm's inventory I decreases each working day as given by $I = 10,000 - 200t$, where t represents days. Graph the inventory equation. After how many days will the stock be depleted?

18 If a boy rides his bicycle at a constant velocity of 4 mph, the distance d (in miles) he will travel in t hours is given by the equation $d = 4t$. Graph the equation.

19 An item sells for $10 under ordinary market conditions. As the supply of the item increases, the price P decreases, as given by $P = 10 - .05x$, where x is the number of units available. Graph the equation.

20 The resistance R in ohms to the flow of electricity through a wire is given by $R = .1x$, where x is the length of the wire in meters. Graph the equation.

21 Let I be the interest, t be the time in years, P be the principal, and r be the rate of interest. These quantities are related by the equation $I = Prt$, the simple interest formula. If $P = \$100$ and $r = 9\%$, then $I = 100(.09)t = 9t$. Graph the equation $I = 9t$.

1.2 Functions

A function is a special relationship that may exist between variables. We will restrict our attention here to functions relating two variables only.

Applied Example 8 A man whose salary is $3/hr knows that his weekly income depends on the number of hours he works in a week.

Note in Table 1.3 that for each man the number of hours worked (represented

TABLE 1.3

	Time x (Hr)	Rule	Salary y ($)
John	40	3(40)	120
Bill	30	3(30)	90
George	15	3(15)	45

by the variable x) produces (by the rule $3 times x) the salary (represented by the variable y) for that man. In this example the salary received by each man is a function of the time worked. Mathematically this is represented by the equation

$$y = 3x$$

In Example 8 the variable x is called the *independent variable*, for it is the variable for which we can choose the value (the number of hours worked). The value of the other variable, y, is dependent on the choice of the value of x (and the rule); therefore y is called the *dependent variable*. If we have no application to guide us, then the independent variable is the one for which we choose values. In graphing $x + y = 4$ (Example 3) we chose values of x and computed the values of y; hence x was independent and y was dependent.

DEFINITION

A function is a correspondence between two variables x (the independent variable) and y (the dependent variable) such that for each value of x there corresponds a single (unique) value of y.

For convenience, the variable y is said to be a function of the variable x, denoted $y = f(x)$. The equation $y = f(x)$ is read "y equals f of x."

Although there are other notations that are used to represent functions, this text will use the notation of the form $y = f(x)$. When a functional relation exists between two variables x and y and the function is represented by an equation, the variable y will often be replaced by $f(x)$.

The use of the letters x and y to represent the variables is strictly arbitrary. Sometimes using other letters such as C for cost, R for revenue, P for profit, t for time, and so on, might be more meaningful. Also the use of f in $f(x)$ is arbitrary (as is x). The symbols $f(t)$, $g(x)$, $h(n)$, and so on, can also be used in functional notation.

Example 9 Let $y = 2x + 3$. In this equation, for each value of x, there corresponds a single (unique) value of y. For example, if $x = 2$, then $y = 7$ only. If $x = 1$, then $y = 5$ only. For any and all values of x, there is only one value of y. Therefore the equation $y = 2x + 3$ represents a function. It is appropriate to replace the dependent variable y with $f(x)$ and rewrite the equation in functional notation as

$$f(x) = 2x + 3$$

When the notation $f(x)$ is used, it is not unusual to encounter expressions such as $f(2)$. The number 2 has replaced x in $f(x)$, and therefore 2 should replace x throughout the equation. Therefore if $f(x) = 2x + 3$, then

$$f(2) = 2(2) + 3 = 7$$
$$f(3) = 2(3) + 3 = 9$$
$$f(-4) = 2(-4) + 3 = -5$$

and in general,

$$f(p) = 2(p) + 3 = 2p + 3$$

There are two special functions that occur in the study of calculus which we would like to mention at this time. The first is called a *constant function*. In this type of function, no matter what value is given to the independent variable, the rule will assign a constant value to the dependent variable.

Applied Example 10 Suppose Harry Hero signs a no-cut contract with a professional football team for a salary of \$30,000 per year. Let the variable t

TABLE 1.4

Time played t (min)	Rule (function)	Salary $f(t)$ (\$)
0	Contract	30,000
30	,,	30,000
200	,,	30,000
t	,,	30,000

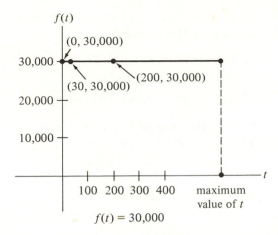

$f(t) = 30,000$

FIGURE 1.11

be the number of minutes that Harry plays during the season. Let his salary be represented by the dependent variable $f(t)$. (See Table 1.4.)

Symbolically this can be represented by the equation $f(t) = 30,000$, for non-negative t up to maximum playing time. Therefore $f(t) = 30,000$ is a constant function. The graph of a constant function is a horizontal line (See Figure 1.11.)

The second special function is one defined by two or more rules, where the rule used to compute the value of the dependent variable depends on the value of the independent variable.

Applied Example 11 A familiar example of a function defined by several rules is the graduated income tax function. Let a person's taxable income be represented by the (independent) variable n. The assigning rule (function) is determined by the amount of the person's taxable income, which is a value of n. The tax to be paid, symbolically $f(n)$, depends on the value of n and on the assigning rule for that specific value of n.

Table 1.5 shows that different rules are used for different values of n. Thus we

TABLE 1.5

Income n ($)	Rule	Tax $f(n)$ ($)
3000	15% of n	450
6000	$600 + 19\%$ of $(n - 4000)$	980
9000	$1360 + 22\%$ of $(n - 8000)$	1580

must have some method of determining which rule is to be used. This type of function can be stated by the following rule:

$$f(n) = \begin{cases} .15n & \text{if } 0 \le n \le 4000 \\ 600 + .19(n - 4000) & \text{if } 4000 < n \le 8000 \\ 1360 + .22(n - 8000) & \text{if } 8000 < n \end{cases}$$

The entries in Table 1.5 were computed in the following manner:

For $n = 3000$, $f(3000) = .15(3000) = 450$, since $n \le 4000$.

For $n = 6000$, $f(6000) = 600 + .19(6000 - 4000) = 980$, since $4000 < n \le 8000$.

For $n = 9000$, $f(9000) = 1360 + .22(9000 - 8000) = 1580$, since $8000 < n$.

The graph of a function defined by several rules is usually not a smooth line. (See Figure 1.12.)

At this point the following question seems natural: Is there a correspondence between two variables that is not a function? Yes. There are many such correspondences. In correspondences that are not functions, the assigning rule is the reason. Nonexample 12 illustrates a correspondence that is not a function.

Nonexample 12

TABLE 1.6

Value of x	Rule	Values of y
4	Square root	-2 and $+2$
9	"	-3 and $+3$
16	"	-4 and $+4$
21	"	$-\sqrt{21}$ and $+\sqrt{21}$
x ($x > 0$)	"	$-\sqrt{x}$ and $+\sqrt{x}$

Note in Table 1.6 that each value of the variable x is not in correspondence with a single (unique) value of the variable y. Therefore the correspondence illustrated by the table is not a function.

The following nonexample will emphasize why functions are more practical in some situations than nonfunctions.

Applied Nonexample 13   The electric company computes Joe's bill based on the number of kilowatt hours that Joe uses in his home each month. Let x be the number of kilowatt hours used for the month of June and y be the bill for the month of June. Suppose that Joe uses 12,000 kwh but the computer at the electric company sends Joe two bills, one for $31.28 and another for $18.17. This provides correspondences of (12,000, $31.28) and (12,000, $18.17). The

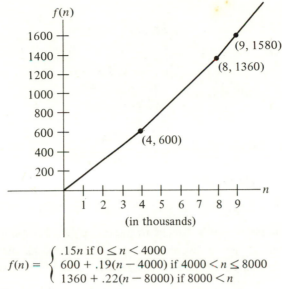

$$f(n) = \begin{cases} .15n \text{ if } 0 \leq n < 4000 \\ 600 + .19(n - 4000) \text{ if } 4000 < n \leq 8000 \\ 1360 + .22(n - 8000) \text{ if } 8000 < n \end{cases}$$

FIGURE 1.12

variable x does not correspond to a single (unique) value of the variable y. Therefore the correspondence is not a function. A company that bills its customers in this manner is certain to receive some complaints.

▶ **Exercise 1.2**

1 Let $y = x + 4$. Is y a function of x?

2 Let $x + y = 6$. Is y a function of x?

3 Let $p = 2^q$. Is p a function of q?

4 Let $f(x) = x + 4$. Find $f(2)$, $f(-3)$, and $f(p)$.

5 Let $f(x) = x^2 - 1$. Find $f(0)$, $f(3)$, and $f(-4)$.

6 Let $f(x) = \frac{1}{2}x$. Find $f(3)$, $f(-4)$, and $f(9)$.

7 Let $C(n) = n + 5$. Find $C(\frac{1}{2})$, $C(0)$, and $C(12)$.

8 Let $R(n) = 3n$. Find $R(\frac{1}{4})$, $R(1.6)$, and $R(3\frac{1}{2})$.

9 Let $p = h(q) = 3q - 5$. Find $h(3)$, $h(0)$, and $h(-5)$.

10 Let

$$f(x) = \begin{cases} x + 3 & \text{if } x \leq 0 \\ x^2 & \text{if } x > 0 \end{cases}$$

Find $f(2)$, $f(0)$, and $f(-3)$.

11 Let

$$g(x) = \begin{cases} 2x + 5 & \text{if } x < 0 \\ 3 & \text{if } x = 0 \\ 1 - x & \text{if } x > 0 \end{cases}$$

Find $g(3)$, $g(\frac{1}{2})$, $g(0)$, and $g(-4)$.

12 Let

$$h(x) = \begin{cases} x - 4 & \text{if } x \leq 2 \\ 5 & \text{if } x > 2 \end{cases}$$

Find $h(0)$, $h(2)$, and $h(4)$.

13 Is the correspondence of parents to their children a function? Explain.

14 Is the correspondence of a student to his test paper on a single test a functional relation? Is the correspondence of a student to all his tests during a semester, if there are more than one test, a functional relation?

15 Is the relation of an arithmetic problem to its answer a function? Explain.

16 A man's income tax will be 19% of his taxable income if his taxable income is $4000 or less. It will be $760 + 22% of the excess over $4000 if his taxable income is between $4000 and $8000, including $8000.
   a Write the function in rule notation.
   b If his taxable income is $5500, find the amount of tax he must pay.
   c If his taxable income is $2000, how much tax must he pay?

17 Jack's contract will pay $1000 when he completes the job.
   a If Jack completes the job in 12 hr, how much will he be paid?
   b If Jack completes the job in 5 yr, how much will he be paid?
   c Write the equation expressing pay as a function of time.

18 The weight of a container plus contents in pounds is a function of the amount of water in the container. If the container weighs 10 lb and water weighs 62.4 lb/cu ft, write an equation relating the total weight w of the container to the number of cubic feet of water n that it contains. What is the total weight of the container if it contains 3 cu ft of water?

19 A firm's production cost is a function of the number of units x produced as given by

$$C(x) = x^2 + 3x + 75$$

Find the cost of producing 10 items.

20 The distance a train travels is a function of the time the train travels. If a train travels at a constant velocity of 50 mph, write an equation representing the distance d traveled in miles as a function of time t in hours. How far will the train travel in 5 hr? How long will it take the train to travel 125 miles?

21 An antibiotic reaches its maximum concentration in the blood immediately after injection. Write an equation representing the concentration c in the blood after t hours if the concentration decreases by 20%/hr. What percentage of the original concentration remains in the blood after 30 min?

1.3 Graphing a Function

In order to graph a function, some solution pairs must be computed. If the pair (p, q) is a solution pair of a function to be graphed, then p is a value of the independent variable and q is the corresponding value of the dependent variable. The value p of the independent variable corresponds to a number on the horizontal axis. The value q of the dependent variable corresponds to a number on the vertical axis. Further, the horizontal axis is named the same as the independent variable and the vertical axis is named the same as the dependent variable.

Example 14 If x is the independent variable and y is the dependent variable, then the horizontal axis is named the x axis and the vertical axis is named the y axis. (See Figure 1.13.)

If the independent variable is n and the dependent variable is $R(n)$, then the horizontal axis is named the n axis and the vertical axis is named the $R(n)$ axis. (See Figure 1.14.)

The need to know exactly what values can be given to the independent variable and what values might be expected for the dependent variable leads us to the following definitions.

FIGURE 1.13

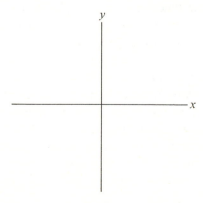

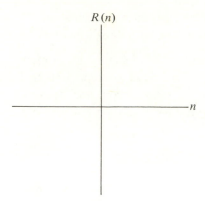

FIGURE 1.14

DEFINITION

The set of values that are to be used for the independent variable is called the *domain of the function* (or simply *domain*).

DEFINITION

The set of all values of the dependent variable that correspond to the values of the independent variable is called the *range of the function* (or simply *range*).

Note: A function whose range is restricted to real numbers is called a real-valued function. From now on, this is the only type of function that will be considered in the text.

In graphing, the domain is important because it is the set of numbers from which we choose the values of the independent variable. Thus in order to graph a function, we must first determine its domain.

In an applied problem the range is of special interest because it is the set of results produced by the function. For example, in a function where the cost depends on the number of units produced, the number of units produced is a value of the domain but the cost of production is a value in the range.

There are three types of restrictions placed on the domain. These restrictions keep us from using some real numbers for the values of the independent variable.

First, there are the mathematical restrictions. These are restrictions such as not allowing the denominator of any fraction to be zero or not allowing a negative factor to occur under any square root symbol.

Second, there are restrictions arising from the application of the function. Suppose a function represents the revenue accrued from the sale of tickets to a football game, where x is the independent variable representing the number of tickets sold. The domain in this case would be only the nonnegative whole numbers 0, 1, 2, 3, ... up to the capacity of the stadium.

Third, there are the arbitrary restrictions. In this case the domain is stated to be a certain set of numbers even though the first two types of restrictions do not apply. The following examples illustrate each of the three types of restrictions.

First consider some mathematical restrictions on the domain. In this case, the most sensible way to proceed is to assume the domain is the set of all real numbers. Then determine if there are any real numbers that must be omitted.

Example 15   Find the domain and graph $y = 1/(x - 3)$, where x is the independent variable.

The denominator $x - 3$ cannot be zero because $1/0$ is not a real number. Since $x - 3$ cannot be zero, $x \neq 3$. To graph, choose some values from the domain that are larger than 3 and some that are smaller than 3. (See Figure 1.15.)

Example 16   Find the domain and graph $y = \sqrt{x - 6}$, where x is the independent variable.

The square root of a negative number is not a real number. Since $x - 6$ must be positive or zero, x must be greater than or equal to 6, that is $(x \geq 6)$. (See Figure 1.16.)

FIGURE 1.15

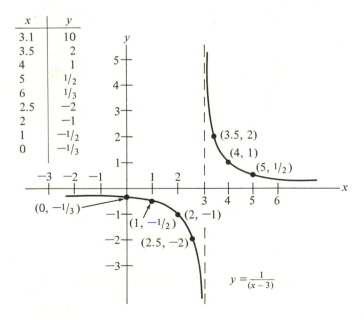

x	y
3.1	10
3.5	2
4	1
5	$1/2$
6	$1/3$
2.5	-2
2	-1
1	$-1/2$
0	$-1/3$

$y = \frac{1}{(x-3)}$

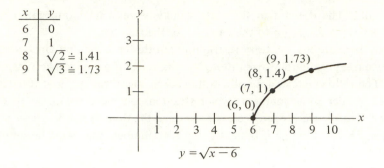

x	y
6	0
7	1
8	$\sqrt{2} \doteq 1.41$
9	$\sqrt{3} \doteq 1.73$

$y = \sqrt{x - 6}$

FIGURE 1.16

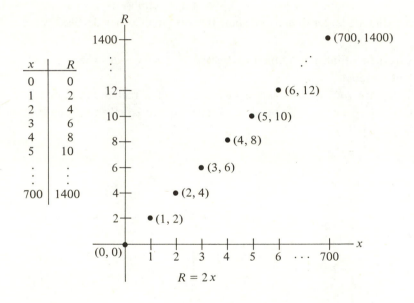

x	R
0	0
1	2
2	4
3	6
4	8
5	10
⋮	⋮
700	1400

$R = 2x$

FIGURE 1.17

The second type of restriction on the domain arises from the application of the function.

Example 17 Find the domain and graph $R = 2x$, where R is the revenue a theater owner can expect from the sale of tickets costing \$2 each. Assume that the theater has 700 seats. Let x be the independent variable. What is the range of the function?

The domain is the set of nonnegative whole numbers 0, 1, 2, 3, 4, ... , 700. The range is the set of numbers 0, 2, 4, 6, ... , 1400.

Note that the graph in Figure 1.17 consists of isolated points, because there are no numbers in the domain between those listed. (Sometimes the points in the graph may be connected with a line in order to make the graph more meaningful even though it is not mathematically correct.)

Example 18   The circumference C of a circle is a function of its diameter d, as given by $C = \pi d$, where π is the nonending, nonrepeating decimal 3.14159. ... (The number 3.14159 ... occurs so frequently in the study of geometry that it is given the symbol designation π for convenience.) For graphing purposes π will be approximated as 3.14. Let d be the independent variable. Find the domain and graph the function.

The domain is the set of nonnegative real numbers. (See Figure 1.18.)

The third type of restriction on the domain is the arbitrary restriction.

Example 19   Graph $y = ex - 2$, where x is the independent variable and the domain is the set of nonnegative real numbers. The letter e is not a variable but represents the nonending, nonrepeating decimal 2.7182. ... (The number 2.7182 ... occurs so frequently in the study of calculus that it, like π, is given a letter designation for convenience. A further discussion of e will be given in some of the following chapters.) For graphing purposes e will be approximated as 2.7. (See Figure 1.19.)

FIGURE 1.18

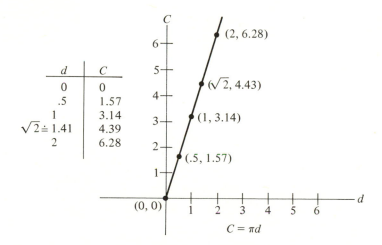

d	C
0	0
.5	1.57
1	3.14
$\sqrt{2} \doteq 1.41$	4.39
2	6.28

$C = \pi d$

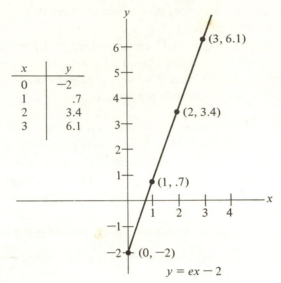

x	y
0	−2
1	.7
2	3.4
3	6.1

$y = ex - 2$

FIGURE 1.19

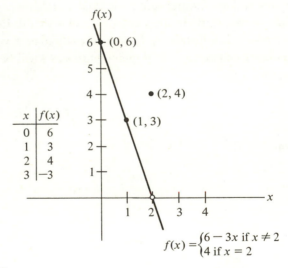

x	f(x)
0	6
1	3
2	4
3	−3

$$f(x) = \begin{cases} 6 - 3x & \text{if } x \neq 2 \\ 4 & \text{if } x = 2 \end{cases}$$

FIGURE 1.20

Example 20 Graph

$$f(x) = \begin{cases} 6 - 3x & \text{if } x \neq 2 \\ 4 & \text{if } x = 2 \end{cases}$$

The independent variable is x. (See Figure 1.20.)

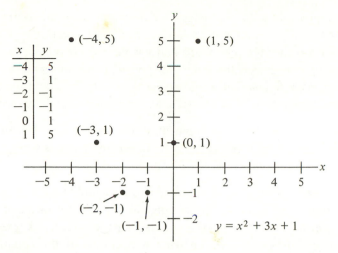

FIGURE 1.21

Example 21   Graph the function $y = x^2 + 3x + 1$, where the domain is -4, $-3, -2, -1, 0, 1$. Let x be the independent variable. Find the range of the function.

The range of the function is $-1, 1, 5$. Note that in Figure 1.21 we do not draw a smooth curve through the points, since the domain consists of only six integers, $-4, -3, -2, -1, 0$, and 1.

▶ **Exercise 1.3**

Find the domain of each of the functions in Problems 1 through 6. Let x be the independent variable. Graph the function in each problem.

1 $y = \dfrac{2}{x - 4}$ **2** $y = \sqrt{x - 5}$

3 $y = \sqrt{2 - x}$ **4** $y = x^2$

5 $y = \dfrac{1}{x + 3}$ **6** $y = 3x - 2$

7 A company knows that its cost of production per day of a certain item is given by $c = 6x + 15$, where x is the number of items produced per day. If the company can produce a maximum of 50 items per day, find the domain and graph the cost as a function of production.

8 A high school has determined an equation for predicting the grade point averages (gpa's) of their graduates who attend the local university by using the students' high school grade point averages. The equation is $y = .60x + .75$, where x is the high school gpa and y is the university gpa. Assume 2.00 on a four-point scale is the minimum gpa that the university will accept. Predict the college gpa that would be earned by students with the following high school gpa's: a. 3.25; b. 2.00; c. 3.80. What is the domain of the function?

9 A car dealer knows that on a shipment of 300 or fewer cars he can make an average profit of $300 per car. Write an equation representing the profit function. Find the domain of the function and draw its graph.

10 An antifreeze company has determined experimentally that a 19-qt capacity cooling system can be kept from freezing, down to a temperature of y degrees Fahrenheit, by the addition of x quarts of its antifreeze, as given by the equation $y = 4x + 14 - x^2$. To what temperature is the cooling system protected from freezing if 3 qt of antifreeze are added? if 10 qt are added? If the domain of the function is from 3 through 11 qt, what is the range?

11 A fraternity is selling tickets to its annual dance for $4 per person. If the profit p earned from the dance is given by $p = 4n - 100$, where n is the number of tickets sold, find the domain of the function. Assume that 100 tickets are available. What is the range of the function? Graph the function.

12 Graph $y = x^2 + 1$, where the domain is 0, 1, 2, 3, and x is the independent variable. Find the range.

13 Graph $y = 7 - x$, where the domain is the set of nonnegative real numbers. Let x be the independent variable. Find the range.

14 Graph

$$g(x) = \begin{cases} x & \text{if } x \text{ is } 1, 2, 3, 4, \ldots \\ 3 & \text{if } x \text{ is } 0 \end{cases}$$

Find the range.

15 In a particular region the velocity of a shock wave passing through the earth at a depth x is $v = .2e^{.5x}$, where v is in kilometers per second (km/sec) and x is in kilometers. Graph the equation for depths from 0 to 4 km. Use the table in Appendix 3 to determine the values of e to an exponent.

16 The amount of radiant energy A absorbed by a particular material depends on the thickness t of the material, according to $A = 10^4 e^{.4t}$. Graph the function. Let the thickness vary from 0 to 5 units.

1.4   The Slope of a Straight Line

DEFINITION

A *linear equation* in two variables, x and y, is an equation that can be expressed
in the form

$$ax + by = c$$

where a, b, and c are constants, a and b not both zero.

Remember that the names of the variables are arbitrary. That is, an equation
of the form $ap + bq = c$, where a, b, and c are constants, a and b not both zero,
is also a linear equation in the two variables p and q.

Example 22   The equations

$$y = 2x + 1$$
$$C = 2n + 10$$
$$3p + 2q = 6$$
$$2x + f(x) = 5$$
$$R(n) = 3n + 12$$

are all linear equations in two variables.

Nonexample 23   The equations

$$y = x^2$$
$$q = 4p - p^2$$
$$y = 2x^3 - 6x + 5$$
$$f(x) = x^2 + 1$$
$$xy = 1$$

are not linear equations in two variables, although they are equations having
two variables.

All linear equations in two variables have graphs that are straight lines.
(See Examples 3, 4, 10, 18, and 19.) Note that each straight line in those examples
has a certain slant or inclination. It is this property we would now like to
consider.

In mathematics the inclination of a straight line is called the *slope* of the line.
The slope is defined as follows.

DEFINITION

Given two different points $(p, f(p))$ and $(q, f(q))$ on a straight line, the *slope* of the line, denoted by m, is the ratio

$$m = \frac{f(p) - f(q)}{p - q}$$

where $p \neq q$.

Remember that two points determine a straight line. Therefore if we have *any* two points on a straight line, the slope of the line can be determined.

Example 24 Find the slope of the straight line through the points $(-5, -2)$ and $(3, 4)$. Let $(p, f(p)) = (-5, -2)$ and $(q, f(q)) = (3, 4)$. Then

$$m = \frac{-2 - 4}{-5 - 3} = \frac{-6}{-8} = \frac{3}{4}$$

It does not matter which point is considered as $(p, f(p))$ or $(q, f(q))$ in the application of the formula. If $(q, f(q)) = (-5, -2)$ and $(p, f(p)) = (3, 4)$, then

$$m = \frac{4 - (-2)}{3 - (-5)} = \frac{6}{8} = \frac{3}{4}$$

Either way, the slope of the line through the points $(-5, -2)$ and $(3, 4)$ is $\frac{3}{4}$.

Example 25 Find the slope of the line passing through the points $(-3, 10)$ and $(1, -2)$ and graph the line. (See Figure 1.22.)
   Let $(p, f(p)) = (-3, 10)$ and $(q, f(q)) = (1, -2)$. Then

$$m = \frac{10 - (-2)}{-3 - 1} = \frac{12}{-4} = -3 \cdot$$

The following examples show some of the applications of the slope of a straight line.

Applied Example 26 Let C be the cost of an item in dollars, r the percentage markup, and S the selling price of the item. These quantities are related by the equation $S = rC + C$. Let $r = 12\%$. Then

$$S = .12C + C$$
$$= C(.12 + 1)$$
$$= 1.12C.$$

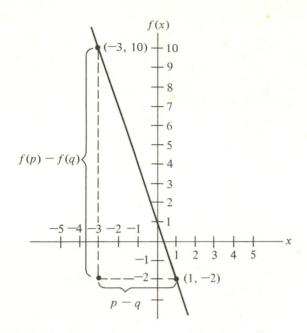

FIGURE 1.22

Graph the selling price as a function of cost, and find the slope of the line.

The slope of the line can be computed by using any two points on the line. Let $(p, f(p)) = (5, 5.60)$ and $(q, f(q)) = (0, 0)$. Then

$$m = \frac{5.60 - 0}{5 - 0} = 1.12$$

Therefore the slope is $1 + .12$ for $S = .12C + C$, and for $S = rC + C = (1 + r)C$ the slope is $1 + r$. (See Figure 1.23.)

Applied Example 27 (Average Velocity) Suppose the distance $s(t)$ in feet an object travels in time t in seconds is $s(t) = t^2 + 20t$.

1. Find the distance the object travels in 5 sec.

$$s(5) = 5^2 + 20(5) = 125 \text{ ft}$$

2. Find the distance the object travels in 10 sec.

$$s(10) = 10^2 + 20(10) = 300 \text{ ft}$$

3. Find the average velocity of the object from $t = 5$ to $t = 10$.

The average velocity of a moving object is the distance the object travels divided by the elapsed time. That is,

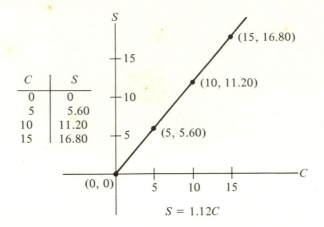

FIGURE 1.23

$$\text{average velocity} = \frac{\text{change in distance}}{\text{change in time}}$$

Therefore the average velocity of the object in this example from $t = 5$ to $t = 10$ is as follows:

$$\text{average velocity} = \frac{300 - 125}{10 - 5} = 35 \text{ ft/sec}$$

Note that the average velocity is the slope of the straight line through the points (10, 300) and (5, 125).

4. Graph distance as a function of time. (See Figure 1.24.)

FIGURE 1.24

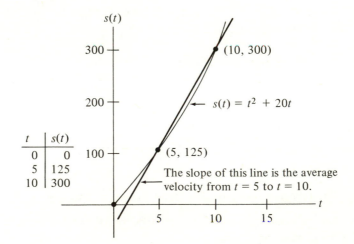

Applied Example 28 (Average Marginal Cost)   The Eastern Radio Company has determined that its cost in dollars of producing its best AM–FM radio is

$$C(x) = \frac{300}{x} + 25x + 1000$$

where x is the number of units produced and $x \geq 4$.

1. Find the cost of producing 10 radios.

$$C(10) = 30 + 250 + 1000 = 1280$$

2. Find the cost of producing 11 radios.

$$C(11) = 27.27 + 275 + 1000 = 1302.27$$

Having computed the cost of producing 10 radios and 11 radios, we can find what economists call the average marginal cost of producing the 11th radio. The average marginal cost is the rate of change of the total cost with respect to the corresponding change in production. That is,

$$\text{average marginal cost} = \frac{\text{change in total cost}}{\text{change in production}}$$

3. Find the average marginal cost of producing the 11th radio.

$$\text{average marginal cost} = \frac{C(11) - C(10)}{11 - 10}$$

$$= \frac{1302.27 - 1280}{1}$$

$$= 22.27$$

FIGURE 1.25

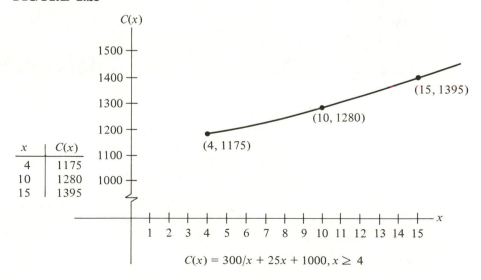

x	$C(x)$
4	1175
10	1280
15	1395

$$C(x) = 300/x + 25x + 1000, x \geq 4$$

Note that the average marginal cost is the slope of the line through the points (10, 1280) and (11, 1302.27).

4. Graph the cost as a function of units produced. (See Figure 1.25.)

▶ **Exercise 1.4**

Tell which of the equations in Problems 1 through 9 are linear. For those that are, find a, b, and c.

1 $y = 4x - 6$ 2 $y = x^3$

3 $q - p = 8$ 4 $\dfrac{2}{x} + \dfrac{3}{y} = 1$

5 $R = 4n + n^2$ 6 $y = \sqrt{x - 5}$

7 $y = \dfrac{1}{x + 3}$ 8 $y = e^x$

9 $y = \pi x - 3$

Find the slope of the line passing through each of the pairs of points in Problems 10 through 16. Plot the two points and draw the line determined by each pair.

10 (1, 1) and (3, 4)   11 (0, 2) and (6, 0)

12 (3, 2) and (3, 5)   13 (−1, 2) and (2, 3)

14 (3, 2) and (5, 2)   15 ($\frac{1}{2}$, −$\frac{3}{2}$) and (3, 4)

16 $(x, f(x))$ and $(p, f(p))$

17 Find the slope of the line $y - 3x = 4$.

18 Find the slope of the line $y = x$.

19 Find the slope of the line $y = 10$.

20 If the cost of producing n items is given by $c = 100 + 5n$, find the slope of the cost function. Find the marginal cost of producing the 10th unit.

21 If an object is dropped vertically, it will fall a distance s ft in t sec as given by $s = 16t^2$ (neglecting air resistance). Find the distance the object will fall in 2 sec. What is the average velocity of the object over the first 2 sec? If the object is dropped from a height of 256 ft, how long will it take to reach the ground?

22 Given the function $g(x) = x^2 + 3x + 5$, find the slope of the line through the points on the graph of the function where $x = 2$ and $x = 5$. The slope of the line is the average rate of change of the function with respect to the change in x from $x = 2$ to $x = 5$.

23 Given the function $y = 16 - x^2$, find the average rate of change of the function with respect to a change in x from $x = 1$ to $x = 3$.

24 If the profit yielded from the sale of x items is represented by $p = 2x + 1/x$, find the marginal profit yielded from the sale of the 10th item. Assume that marginal profit is defined similarly to marginal cost.

25 If the cost c is related to the number of units x produced by $c = 100 + 2x + \sqrt{x}$, find the marginal cost of producing the 16th unit. (Let $\sqrt{15} = 3.87$.)

26 A psychologist tested his theory of teaching reading on two groups of children for 12 months. In the experimental group he found that the average reading level increased from a level of 2.8 to a level of 5.2. Find the average increase per month in reading level for this group. In the traditional group he found that the average reading level increased from a 2.8 level to a 3.6 level. Find the average increase per month in reading level for the traditional group.

1.5   The Equation of a Straight Line

Recall that all linear equations have graphs that are straight lines. By using the expression for the slope of a line, a formula can be developed that will allow us to write the equation of a straight line. To find this equation we must know the slope of the line and one point on the line.

Let (p, q) be a fixed point on the line and (x, y) be a general point on the line distinct from (p, q). Then the slope of the line is

$$m = \frac{y - q}{x - p} \qquad (x \neq p)$$

Multiplying by $x - p$ yields

$$(x - p)m = y - q$$

the *point-slope formula* of a straight line. It is so named because it is the equation of the straight line in variables x and y, with slope m and passing through the point (p, q).

Example 29   Find the equation of the line that passes through the point $(1, 3)$ and has a slope of 2.

Let $(p, q) = (1, 3)$ and $m = 2$. Then $y - 3 = 2(x - 1)$. Simplifying, $y = 2x + 1$.

Example 30   Find the equation of the line that passes through the point $(-4, -2)$ and has a slope of $-\frac{1}{2}$.

Let $(p, q) = (-4, -2)$ and $m = -\frac{1}{2}$. Then

$$y - (-2) = -\tfrac{1}{2}[x - (-4)]$$

Simplifying,

$$y + 2 = -\tfrac{1}{2}(x + 4)$$
$$= -\tfrac{1}{2}x - 2$$

Multiplying by 2,

$$2y + 4 = -x - 4$$

Hence,

$$x + 2y = -8$$

Example 31 Find the equation of the line that has a y intercept of 3 and a slope of 7.

A y intercept occurs at $x = 0$; therefore the point $(p, q) = (0, 3)$ and $m = 7$. Then $y - 3 = 7(x - 0)$. Simplifying, $7x - y = -3$.

Applied Example 32 Let n be the number of units of a certain item in stock and d be days. At the beginning of the month (at zero days) the store had 500 units of a certain item in stock. After 15 days it had 200 units in stock.

1. Write an equation relating the number of units in stock to the day of the month, assuming the stock depletion is linear.

By using the data given, the slope

$$m = \frac{500 - 200}{0 - 15} = -20$$

With $m = -20$ and the point $(0, 500)$, the point-slope formula yields $n - 500 = -20(d - 0)$. Simplifying, $n = -20d + 500$.

2. How many units are in stock after five days have passed? Let $d = 5$; then $n = -20(5) + 500 = 400$ units.

3. After how many days will the stock be depleted? Let $n = 0$; then $0 = -20d + 500$. Solving, $d = 25$ days.

▶ **Exercise 1.5**

Write the equation of the line that satisfies each of the conditions in Problems 1 through 8 (let the variables be x and y).

1 Passes through $(2, 3)$ and has slope 5

2 Passes through $(1, \tfrac{1}{2})$ and has slope -2

3 Passes through $(1, 3)$ and $(7, 8)$

4 Passes through $(-2, -2)$ and $(1, 2)$

5 Passes through $(3, 2)$ and $(5, 2)$

6 Passes through the origin and $(-1, 2)$

7 Has slope of -5 and y intercept of 2

8 Has slope of 0 and passes through $(1, 1)$

9 Let k be the cost of an item and s be its selling price.
   a Write an equation relating the selling price to the cost when the rate of markup is 15%. Assume there is a linear relationship between s and k.
   b What is the selling price of an item that costs $30?
   c What is the cost of an item that sells for $38.75?

10 In a farmer's first year he produced 8000 bushels of wheat. In his second year he produced 8200 bushels of wheat. If the farmer can continue to increase his production at the same rate, write an equation relating production to years. How much wheat will the farmer produce in his seventh year?

11 Suppose a person who weighs 120 lb on earth weighs 96 lb on Venus, while a person who weighs 100 lb on earth weighs 80 lb on Venus. Assuming a linear relation, write an equation relating earth weight to Venusian weight. How much would a person who weighs 189 lb on earth weigh on Venus? (Let w represent earth weight and v represent Venusian weight.)

12 The 1968 records of the Doomed Co. show that 8000 units produced and sold gave a profit of $45,000. The 1972 records show that 10,000 units produced and sold gave a profit of $25,000. Find the average rate of change of profit with respect to units produced and sold. Write the equation relating profit to the number of units sold (assume linear).

13 Suppose a company stocks 1000 units of an item. After 50 days the stock is depleted. If the depletion of the stock is linear, write an equation relating the number of items in stock to the number of days.

14 A scale is to be made from a coil spring as shown in Figure 1.26. If 1 oz is placed in the container, the spring is elongated .7 in. If 5 oz are placed in the container, the spring is elongated 1.5 in. If the elongation is linearly related to the weight (and the limit of elasticity is not exceeded), write an

FIGURE 1.26

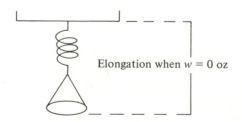

Elongation when $w = 0$ oz

equation relating the elongation to the weight. What would be the elongation if 10 oz were placed in the container? If the spring is elongated 1 in., what is the weight in the container?

15 The revenue from the sale of nine bicycles is $594. If the revenue from the sale of 10 bicycles is $660, find the average change in revenue. Find the equation of the revenue function (assume linear).

16 A psychological testing service gives an aptitude test for a large company. The test is scored on a scale of 0 to 25. The company then records the number of units of work that the employee can do. The testing service wanted to use the aptitude score to predict an employee's productivity. They took a random sample of 10 employees; their two sets of scores are shown in Table 1.7.

TABLE 1.7

Aptitude score	Units of work
10	19
12	25
15	29
18	38
22	43
6	21
10	24
8	18
15	37
4	7

a Plot the points with the aptitude score corresponding to the x axis. (Note that the points are not on the same line.)

b By statistical methods, a line can be found for these data that "best fits" the data. This line is called a *regression line*. If the slope of the regression line is determined to be 1.856 and the y intercept is 3.828, find the equation of the regression line.

c Using the equation found in (b), find how many units of work could be expected of an employee who scored 20 on the aptitude test.

17 Let u be the number of units produced by a company, and $C(u)$ the cost of this production. A spot check shows that the cost of producing 1500 units is $1200. If the average rate of increase in cost with respect to production is .75, write an equation relating cost to production (assume linear). What is the cost of producing 1000 units?

self-test • chapter one

1 Let $f(x) = x^2 + 3$ Find $f(0)$, $f(3)$, and $f(-2)$.

2 Let

$$f(x) = \begin{cases} 2x + 3 & \text{if } x \geq 5 \\ 5 & \text{if } 1 < x < 5 \\ x & \text{if } x \leq 1 \end{cases}$$

Find $f(3)$, $f(6)$, and $f(1)$.

3 A club is selling tickets for a chili supper for $2.50 per person. If the profit for the chili supper is given by $P = 2.50n - 400$, find the profit when 400 tickets are sold. How many tickets must be sold in order for the club to break even? If 1000 tickets are printed, what is the domain of the function? What is the range of the function?

4 Find the slope of the line that passes through the points (5, 7) and (2, −3).

5 Find the equation of the line that has a slope of $\frac{3}{2}$ and passes through the point (4, 6).

6 Graph $y = 3x + 2$.

7 Find the mathematical restrictions on the domain of each of the following.

  a $y = \sqrt{100 - x}$ **b** $y = 1/(3x + 4)$

chapter two

limits

2.1 The Concept of the Limit of a Function

In the previous chapter it was demonstrated that in some functions a functional value could not be obtained for every real number. This lack of a functional value occurred when a particular real number was not in the domain of the function, either when the domain was arbitrarily restricted or when the function was undefined for that particular value. Consider $f(x) = 1/x$ at $x = 0$. No functional value exists at $x = 0$, since $f(0)$ is undefined. Also consider $f(x) = (x^2 - 5x + 6)/(x - 2)$ at $x = 2$. Since $f(2) = 0/0$, no functional value exists at $x = 2$.

But even though no functional value exists at the point in question, we would like to be able to determine the behavior of the function for values near the point. This leads us to the concept of the limit of a function. The limit of a function is most useful when the value of the function cannot be determined.

The idea is to examine the behavior of the functional values that are obtained from values of x that are close to a fixed value p. If $f(p)$ does not exist, the limit of a function must be used to examine the behavior of the function as x approaches p. The concept of the limit of a function, sometimes called simply the *limit*, will be investigated in the following sections.

First we must examine values of the dependent variable. Do the values of the dependent variable approach a finite value* as the values of the independent variable approach a fixed number? If the answer is yes, then the limit of the function exists, and the limit is the number that the values of the dependent variable are approaching. If the values of the dependent variable do not approach

---

* The term *finite value* (though redundant) is often used for emphasis instead of the usual *real number*.

a finite value, then we say that the function has no limit or that the limit does not exist.

Example 1   Let

$$f(x) = \frac{x^2 + x - 2}{x - 1}$$

[Note that $f(1) = 0/0$.] To investigate the behavior of the function near 1, we must find the limit of the function as x approaches 1. We choose some values of x approaching 1 and compute the corresponding values of $f(x)$ as shown in Table 2.1.

TABLE 2.1

Values of $x < 1$	Values of $f(x)$	Values of $x > 1$	Values of $f(x)$
.5	2.5	1.05	3.05
.9	2.9	1.01	3.01
.99	2.99	1.001	3.001
.995	2.995	1.0005	3.0005
.9999	2.9999	1.00001	3.00001

As x approaches 1, the values of $f(x)$ approach 3. Therefore, we say the limit of the function $f(x) = (x^2 + x - 2)/(x - 1)$ is 3, as x approaches 1.

Before defining the limit of a function, it is convenient to introduce the following symbols:

1 $f(x)$ will be used to represent the functional values.

2 p will be used to represent the fixed value that the values of the independent variable x are approaching.

3 $\rightarrow$ will be used for the word *approaches* and $x \rightarrow p$ will be read x *approaches* p.

4 L will be used to represent the limit of the function, that is, the value the dependent variable approaches as $x \rightarrow p$.

DEFINITION

If the values of the dependent variable, $f(x)$, approach a finite number L as the values of the independent variable approach a fixed number p, then L is said to be the limit of the function as x approaches p. This is denoted by

$$\lim_{x \to p} f(x) = L$$

The following facts are implied in the definition of a limit:

1 The values of x must approach p from both the left side and the right side. (See Table 2.1.)

2 The values of x do not equal the value of p. In fact, the value of p may not be in the domain of the function. (See Example 1.)

3 The values of $f(x)$ can be made as close as you please to L.

4 The choices of the values of x that approach p are strictly arbitrary.

It should be observed that the definition of the limit of a function does not tell us how to find the limit of a function. Thus some method must be developed that will actually find the limit, if it exists. The method based on the definition developed in this section is a tabular approach. Values of the independent variable near p are chosen, with each successive choice being closer to p. Then their functional values are computed. These results are then analyzed. If the functional values approach a finite number, then the limit exists, and the value of the limit is that finite number.

Example 2 deals with the function used in Example 1, that is, $f(x) = (x^2 + x - 2)/(x - 1)$. The purpose of Example 2 is to illustrate statement 4 following the definition of a limit by showing another way the values of x may approach 1.

Example 2

TABLE 2.2

Values of $x < 1$	Values of $f(x)$	Values of $x > 1$	Values of $f(x)$
2/3	2.666666 ...	1.2	3.2
7/8	2.875	1.04	3.04
54/55	2.981818 ...	1.008	3.008
124/125	2.992	451/450	3.00222 ...
749/750	2.99866 ...	801/800	3.000125

As x approaches 1, the values of $f(x)$ approach 3. (See Table 2.2.) Therefore $\lim_{x \to 1} f(x) = 3$. The values of x that approach 1 were chosen in an arbitrary manner. In fact, any choice of numbers that approach 1 in this example must yield values of $f(x)$ that approach 3 to make true the equation $\lim_{x \to 1} (x^2 + x - 2)/(x - 1) = 3$.

Example 3   Let $f(x) = 2x$. Find the limit of the function as x approaches 2.

TABLE 2.3

Values of $x < 2$	Values of $f(x)$	Values of $x > 2$	Values of $f(x)$
1.8	3.6	2.2	4.4
1.9	3.8	2.1	4.2
1.99	3.98	2.01	4.02
1.999	3.998	2.001	4.002
1.99995	3.99990	2.00005	4.00010

As the values of x approach 2 from both sides, the values of $f(x)$ approach 4. (See Table 2.3.) Therefore the limit of the function is 4 as $x \rightarrow 2$. Note that, in this case, 2 is in the domain of the function.

Example 4   Let $f(x) = (x^2 - 5x + 6)/(x - 3)$. Find the limit of the function as $x \rightarrow 3$. Note that $f(3) = 0/0$, so 3 is not in the domain of the function. Symbolically, this result is written as $\lim_{x \rightarrow 3} (x^2 - 5x + 6)/(x - 3) = 1$.

TABLE 2.4

Values of $x < 3$	Values of $f(x)$	Values of $x > 3$	Values of $f(x)$
2.9	.9	3.01	1.01
2.95	.95	3.0005	1.0005
2.999	.999	3.00001	1.00001
2.999995	.999995	3.0000001	1.0000001

Computing the values of $f(x)$ for Table 2.4 can be done easily by algebraically simplifying the fraction. Thus $f(x) = (x^2 - 5x + 6)/(x - 3) = [(x - 3)(x - 2)]/(x - 3) = x - 2$, $x \neq 3$. This is to say, as long as $x \neq 3$, $x - 3 \neq 0$, and so the $x - 3$ factor in the numerator and the denominator can be canceled.

Example 5   Let

$$f(x) = \begin{cases} x^2 & \text{if } x \neq 1 \\ 5 & \text{if } x = 1 \end{cases}$$

Find the limit of the function as x approaches 1. Note that 1 is in the domain of the function.

TABLE 2.5

Values of x < 1	Values of f(x)	Values of x > 1	Values of f(x)
.9	.81	1.1	1.21
.99	.9801	1.01	1.0201
.9995	.99900025	1.0005	1.00100025
.99998	.9999600004	1.00002	1.0000400004

As may be seen from Table 2.5, $\lim_{x\to 1} f(x) = 1$. In this case, $f(1) = 5$ and $\lim_{x\to 1} f(x) = 1$. Consequently $\lim_{x\to 1} f(x) \neq f(1)$.

▶ **Exercise 2.1**

1 Let $f(x) = (x^2 - 6x + 8)/(x - 4)$. Find the limit of the function as $x \to 4$. Complete Table 2.6 by computing $f(x)$ for each of the given values of x. (*Hint:* First factor the numerator and reduce the fraction; then compute the values of $f(x)$.)

TABLE 2.6

Values of x > 4	Values of f(x)	Values of x < 4	Values of f(x)
4.1		3.9	
4.01		3.99	
4.005		3.9995	
4.00001		3.99999	

2 Let $f(x) = (x^2 - 9)/(x + 3)$. Find the limit of the function as $x \to -3$. Complete Table 2.7 by computing $f(x)$ for each of the given values of x. Is -3 in the domain of the function?

TABLE 2.7

Values of x > −3	Values of f(x)	Values of x < −3	Values of f(x)
−2.5		−3.5	
−2.95		−3.02	
−2.998		−3.002	
−2.99875		−3.00002	

3 Let $f(x) = x^2 + 1$. Find the limit of the function as $x \to 1$. Complete Table 2.8 by computing $f(x)$ for each of the given values of x. Is 1 in the domain of the function?

TABLE 2.8

Values of $x < 1$	Values of $f(x)$	Values of $x > 1$	Values of $f(x)$
1/2		3/2	
3/4		5/4	
24/25		26/25	
99/100		101/100	

4 In Problem 1 make another table in which the values of x approach 4 but are different from those given.

5 In Problem 2 make another table in which the values of x approach -3 but are different from those given.

6 In Problem 3 make another table in which the values of x approach 1 but are different from those given.

7 Let $f(x) = (x^2 - 8x + 7)/(x - 7)$. Find the limit of the function as $x \to 7$. Make your own table. Is $x = 7$ in the domain of the function?

8 Let $f(x) = 2x - 3$. Find the limit of the function as $x \to 0$. Make your own table. Is $x = 0$ in the domain of the function?

9 Let $f(x) = (2x^2 - 7x + 6)/(x - 2)$. Find the limit of the function as $x \to 2$. Make your own table. Is $x = 2$ in the domain of the function?

10 Let

$$f(x) = \begin{cases} x & \text{if } x \geq 0 \\ -x & \text{if } x < 0 \end{cases}$$

Find the limit of the function as $x \to 0$. Is $x = 0$ in the domain of the function?

11 Let

$$f(x) = \begin{cases} 2x + 1 & \text{if } x \neq 1 \\ 2 & \text{if } x = 1 \end{cases}$$

Find the limit of the function as $x \to 1$. Is $x = 1$ in the domain of the function?

12 Let

$$f(x) = \begin{cases} x + 4 & \text{if } x > 3 \\ 2x + 1 & \text{if } x \le 3 \end{cases}$$

Find the limit of the function as $x \to 3$. Is $x = 3$ in the domain of the function?

2.2 Nonexamples of Limits

In the previous section some examples were presented in which the limit of the function existed. The purpose of this section will be to present some nonexamples of limits; that is, situations in which the functional values do not approach a single finite number as x approaches p.

Nonexample 6 Let $f(x) = 1/x^2$. Find the limit of the function as x approaches zero. (Zero is not in the domain of the function.)

TABLE 2.9

Values of $x > 0$	Values of $f(x)$	Values of $x < 0$	Values of $f(x)$
1	1	−1	1
1/2	4	−1/2	4
1/3	9	−1/3	9
1/5	25	−1/5	25
1/100	10,000	−1/100	10,000
1/10,000	100,000,000	−1/10,000	100,000,000

As x approaches zero, the values of $f(x)$ get larger and larger. (See Figure 2.1.) Since there is no largest number in the real number system, the values of $f(x)$ do not approach a finite number. Hence, the limit of this function does not exist as x approaches zero.

When values get larger and larger, they are said to increase without bound. When this happens the functional values are considered to approach infinity, symbolized ∞. We can then symbolize the fact that the function increases without bound, that is, approaches infinity, as $x \to 0$ by the following statement:

$$\lim_{x \to 0} \frac{1}{x^2} = \infty$$

Nonexample 7 Let

$$f(x) = \begin{cases} x + 6 & \text{if } x > 1 \\ x^2 & \text{if } x \le 1 \end{cases}$$

Find the limit of the function as x approaches 1. Note that 1 is in the domain of the function.

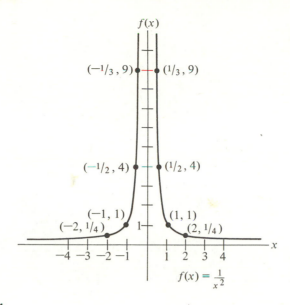

FIGURE 2.1

TABLE 2.10

Values of x > 1	Values of f(x)	Values of x < 1	Values of f(x)
1.5	7.5	.5	.25
1.01	7.01	.99	.9801
1.0001	7.0001	.9995	.99900025
1.000005	7.000005	.999991	.99999888881

This limit does not exist, since the values of the function approach two dif-ferent numbers (1 and 7) depending on whether x approaches 1 from the left side or from the right side. (See Table 2.10 and Figure 2.2.)

It is sometimes useful to note that the values of $f(x)$ approach a finite number as the values of x approach p from one side only. This is called a *one-sided limit*. A one-sided limit is called a *right-hand limit* if all the values of x that approach p are to the right of p on the number line. A one-sided limit is called a *left-hand limit* if all the values of x that approach p are to the left of p on the number line.

In Nonexample 7 the right-hand limit of the function is 7. That is, as $x \rightarrow 1$ from the right side only, the values of $f(x)$ approach 7. The left-hand limit of the function is 1. That is, as $x \rightarrow 1$ from the left side only, the values of $f(x)$ approach 1. Therefore, there is a left-hand limit as $x \rightarrow 1$ and there is a right-hand limit as $x \rightarrow 1$ for the function in Nonexample 7. Yet the limit of the function as

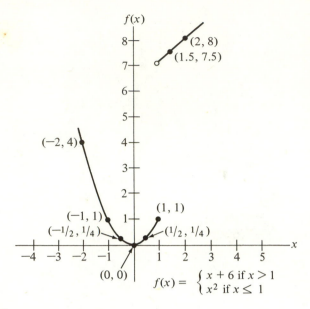

$$f(x) = \begin{cases} x + 6 \text{ if } x > 1 \\ x^2 \text{ if } x \leq 1 \end{cases}$$

FIGURE 2.2

$x \to 1$ does not exist. If the limit of a function exists as $x \to p$, then the right-hand limit must equal the left-hand limit.

▶ **Exercise 2.2**

Investigate each of the following functions in order to determine if the limit exists.

1 Let $f(x) = 1/x$. Find the limit of the function as $x \to 0$.

2 Let

$$f(x) = \begin{cases} x + 3 & \text{if } x > 4 \\ x & \text{if } x \leq 4 \end{cases}$$

  a Find the right-hand limit as $x \to 4$.
  b Find the left-hand limit as $x \to 4$.
  c Does the limit of the function exist as $x \to 4$?

3 Let $f(x) = (x - 1)/(x - 2)$. Find the limit of the function as $x \to 2$.

4 Let

$$f(x) = \begin{cases} x^2 & \text{if } x \geq 2 \\ 4 & \text{if } x < 2 \end{cases}$$

  a Find the right-hand limit as $x \to 2$.

b Find the left-hand limit as $x \to 2$.

c Does the limit of the function exist as $x \to 2$?

5 Let $f(x) = \sqrt{x}$. Find the right-hand limit as $x \to 0$.

2.3   Techniques for Finding Limits as $x \to p$

It seems natural to ask, Is there a way to find the limit of a function without computing a table of values each time? Yes, there is. In this section techniques will be introduced that simplify the problem of finding the limit of a function.

Consider $\lim_{x \to p} f(x)$, where p is a finite value. First, compute $f(p)$. Either $f(p)$ is a finite value or $f(p)$ is not a finite value. If $f(p)$ is a finite value, the following situations occur. The value of $f(p)$ is often the limit of $f(x)$ as $x \to p$. But in our context there are exceptions that must be noted. First, the domain of the function may be arbitrarily restricted. Second, the function may be defined by two or more rules. If either of these possibilities occur, the existence of a limit must be examined more carefully, perhaps by the tabular method.

If $f(p)$ is not a finite value, then consider the following approach: When a function is undefined for a particular value of x, such as p, there are mathematical restrictions on the domain of the function. We are concerned with one particular restriction. That restriction occurs for values of x that make the denominator of the function equal to zero. There are two such cases illustrated in the following examples.

Example 8   Let $f(x) = (x^2 - 5x + 6)/(x - 3)$. At $x = 3$, $f(3) = (9 - 15 + 6)/(3 - 3) = 0/0$. This result does not produce a functional value at $x = 3$.

Example 9   Let $f(x) = (x^2 - 6x + 5)/(x - 3)$. At $x = 3$, $f(3) = (9 - 18 + 5)/(3 - 3) = -4/0$. This result is undefined.

On the surface it might appear that the results in the Examples 8 and 9 are the same. Both of them fail to produce a value for $f(3)$. Yet, in searching for the $\lim_{x \to p} f(x)$, $f(p) = 0/0$ tells us to continue our search for the limit but to use a different attack. Algebra, usually factoring, may be used in order to simplify the function. Then the value of the limit can sometimes be found by evaluating the simplified function at $x = p$.

Example 10   Let $f(x) = (x^2 - 5x + 6)/(x - 3)$. Find the limit of the function as $x \to 3$. Since $f(3) = 0/0$, the following procedure can be used:

$$\lim_{x \to 3} \frac{x^2 - 5x + 6}{x - 3} = \lim_{x \to 3} \frac{(x - 3)(x - 2)}{x - 3} = \lim_{x \to 3} (x - 2) = 1$$

The reduction of $(x - 3)/(x - 3) = 1$ is perfectly legal, since $x \to 3$ implies that $x \neq 3$. Hence, $(x - 3)/(x - 3)$ cannot be $0/0$.

Example 11 Let

$$f(x) = \frac{1/(2 + x) - 1/2}{x}$$

Find the limit of the function as $x \to 0$, if possible.

$$\lim_{x \to 0} \frac{1/(2 + x) - 1/2}{x} = \lim_{x \to 0} \frac{(2 - 2 - x)/(2 + x)(2)}{x}$$

$$= \lim_{x \to 0} \frac{-x}{(2 + x)(2)(x)}$$

$$= \lim_{x \to 0} \frac{-1}{(2 + x)(2)} = -\frac{1}{4}$$

Example 12 Let $f(x) = (x^3 - 64)/(x - 4)$. Find the limit of the function as $x \to 4$, if possible. Since $x^3 - 64 = (x - 4)(x^2 + 4x + 16)$, we have

$$\lim_{x \to 4} \frac{x^3 - 64}{x - 4} = \lim_{x \to 4} \frac{(x - 4)(x^2 + 4x + 16)}{x - 4}$$

$$= \lim_{x \to 4} (x^2 + 4x + 16) = 48$$

Example 13 Let $f(x) = (2 - \sqrt{4 - x})/x$. Find the limit of the function as $x \to 0$. The algebraic simplification that is needed in this problem involves multiplying by 1, where 1 is in the form of $(2 + \sqrt{4 - x})/(2 + \sqrt{4 - x})$.

$$\lim_{x \to 0} \frac{2 - \sqrt{4 - x}}{x} = \lim_{x \to 0} \frac{2 - \sqrt{4 - x}}{x} \cdot \frac{2 + \sqrt{4 - x}}{2 + \sqrt{4 - x}}$$

$$= \lim_{x \to 0} \frac{4 - (4 - x)}{x(2 + \sqrt{4 - x})}$$

$$= \lim_{x \to 0} \frac{x}{x(2 + \sqrt{4 - x})}$$

$$= \lim_{x \to 0} \frac{1}{2 + \sqrt{4 - x}}$$

$$= \frac{1}{2 + \sqrt{4 - 0}} = \frac{1}{4}$$

Recall that $f(3) = -4/0$ in Example 9, where $f(x) = (x^2 - 6x + 5)/(x - 3)$. As $x \to 3$, the limit of that function does not exist, since no algebraic simplifications will help in this problem.

In many situations it is useful to have some limit theorems. These theorems enable us to manipulate the problem into a more useful form. They will be stated here without proof. Then they can be used when needed in more abstract or difficult cases.

LIMIT THEOREMS

If $\lim_{x \to p} f(x) = L$ and $\lim_{x \to p} g(x) = M$ are two finite values, then

1 $\displaystyle\lim_{x \to p} c = c$, where c is a constant

2 $\displaystyle\lim_{x \to p} c \cdot f(x) = c \lim_{x \to p} f(x) = c \cdot L$

3 $\displaystyle\lim_{x \to p} [f(x) \pm g(x)] = \lim_{x \to p} f(x) \pm \lim_{x \to p} g(x) = L \pm M$

4 $\displaystyle\lim_{x \to p} [f(x) \cdot g(x)] = \lim_{x \to p} f(x) \cdot \lim_{x \to p} g(x) = L \cdot M$

5 $\displaystyle\lim_{x \to p} \frac{f(x)}{g(x)} = \frac{\lim_{x \to p} f(x)}{\lim_{x \to p} g(x)} = \frac{L}{M}$ provided that $\lim_{x \to p} g(x) = M \neq 0$

6 $\displaystyle\lim_{x \to p} [f(x)]^n = \left[\lim_{x \to p} f(x)\right]^n = L^n$

Example 14   Let $f(x) = (x^2 + 3x)/(x^3 - 1)$. Find the limit of the function as $x \to 2$. Use the limit theorems.

By Theorem 5,

$$\lim_{x \to 2} \frac{x^2 + 3x}{x^3 - 1} = \lim_{x \to 2} (x^2 + 3x) \div \lim_{x \to 2} (x^3 - 1)$$

By Theorem 3,

$$\lim_{x \to 2} (x^2 + 3x) \div \lim_{x \to 2} (x^3 - 1) = \left(\lim_{x \to 2} x^2 + \lim_{x \to 2} 3x\right) \div \left(\lim_{x \to 2} x^3 - \lim_{x \to 2} 1\right)$$

By Theorem 6,

$$\left(\lim_{x \to 2} x^2 + \lim_{x \to 2} 3x\right) \div \left(\lim_{x \to 2} x^3 - \lim_{x \to 2} 1\right)$$

$$= \left[\left(\lim_{x \to 2} x\right)^2 + \lim_{x \to 2} 3x\right] \div \left[\left(\lim_{x \to 2} x\right)^3 - \lim_{x \to 2} 1\right]$$

By Theorem 1 and Theorem 2,

$$\left[\left(\lim_{x \to 2} x\right)^2 + \left(\lim_{x \to 2} 3x\right)\right] \div \left[\left(\lim_{x \to 2} x\right)^3 - \lim_{x \to 2} 1\right]$$

$$= [2^2 + 3(2)] \div (2^3 - 1) = 10/7$$

▶ **Exercise 2.3**

Find the limits in Problems 1 through 12. Use the techniques of algebraic simplification whenever necessary.

1 $\displaystyle\lim_{x \to 1} (2x + 5)$

2 $\displaystyle\lim_{x \to 1} (x^2 - x - 1)$

3 $\displaystyle\lim_{x \to 0} \frac{x^2 + 2x}{x}$

4 $\lim\limits_{p \to 12} \sqrt{p} - 8$ ($\sqrt{}$ is understood to be only the positive square root)

5 $\lim\limits_{x \to -2} \dfrac{(x+1)(x+2)}{x+2}$

6 $\lim\limits_{x \to 8} \sqrt[3]{x}$

7 $\lim\limits_{x \to 2} x - \dfrac{1}{x}$

8 $\lim\limits_{x \to 2} \dfrac{x-2}{x}$

9 $\lim\limits_{x \to 2} \dfrac{x^3 - 8}{x - 2}$

10 $\lim\limits_{x \to 3} \dfrac{x^2 + 2x}{2}$

11 $\lim\limits_{x \to 0} \dfrac{3 - \sqrt{9-x}}{x}$ [*Hint:* Multiply by $(3 + \sqrt{9-x})/(3 + \sqrt{9-x}) = 1$.]

12 $\lim\limits_{x \to 0} \dfrac{5 - \sqrt{25-x}}{x}$

13 Let

$$f(x) = \begin{cases} x^2 & \text{if } x \ge 0 \\ 2x & \text{if } x < 0 \end{cases}$$

Find the limit of the function as $x \to 0$.

14 Let

$$f(x) = \begin{cases} x^3 & \text{if } x \ge 0 \\ 3x + 5 & \text{if } x < 0 \end{cases}$$

Show that this function has no limit as $x \to 0$.

FIGURE 2.3

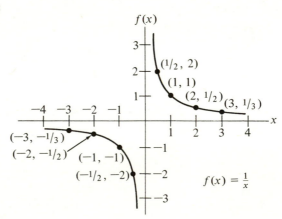

2.4 Limits as $x \to \infty$

In this section we will emphasize limits in which x does not approach a fixed number. We will allow x to increase without bound in the positive direction, denoted by $x \to \infty$, or in the negative direction, denoted by $x \to -\infty$. Since x does not approach a finite value, a new definition must be formulated.

DEFINITION

If the values of the dependent variable, $f(x)$, approach a single number L as x increases without bound in the positive direction, then L is the limit of the function as $x \to \infty$. This is denoted by

$$\lim_{x \to \infty} f(x) = L$$

Example 15 Let $f(x) = 1/x$. Find the limit of the function as $x \to \infty$.

TABLE 2.11

Values of x	Values of $f(x)$
100	$1/100 = .01$
5,000	$1/5000 = .0002$
150,000	$1/150,000 = .00000667$
1,000,000	$1/1,000,000 = .0000001$

From Table 2.11 we can see that the values of $f(x)$ approach zero as the values of x increase without bound. Therefore the $\lim_{x \to \infty} (1/x) = 0$. (See Figure 2.3.)

Example 16 Let $f(x) = (x + 1)/(2x - 1)$. Find the limit of the function as $x \to \infty$.

TABLE 2.12

Values of x	Values of $f(x)$
10	$11/19 \doteq .5789$
100	$101/199 \doteq .5075$
5,000	$5,001/9,999 \doteq .5001$
1,000,000	$1,000,001/1,999,999 \doteq .50000075$

Thus $\lim_{x \to \infty} [(x + 1)/(2x - 1)] = \frac{1}{2}$. (See Table 2.12 and Figure 2.4.)

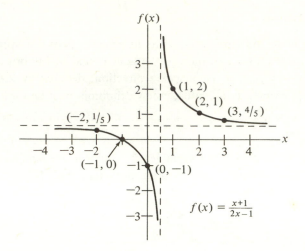

FIGURE 2.4

Nonexample 17   Let $f(x) = (x^2 + 1)/x$. Find the limit of the function as $x \to \infty$.

TABLE 2.13

Values of x	Values of $f(x)$
10	10.1
50	50.02
1000	1000.001

Since the values of $f(x)$ do not approach a finite value as $x \to \infty$, the limit of this function does not exist. (See Table 2.13 and Figure 2.5.)

Example 18   Let $f(x) = (1 + 1/x)^x$. Using Table 2.14, we investigate the behavior of the functional values as $x \to \infty$. The sets of three dots in the table mean that the decimals are infinite. The values in Table 2.14 have been rounded to five places.

As $x \to \infty$, the values of $f(x)$ approach $2.71828 \ldots$, which is a nonrepeating, nonending decimal. Since it would be impossible to write this decimal in full, and this limit turns out to be very important, the symbol e has been selected to represent the constant value that the function approaches. (See Example 19, Chapter 1.) Thus, $\lim_{x \to \infty} (1 + 1/x)^x = e$. The symbol e will be referred to many times in the text. As the properties of e are discussed, the great number of applications involving this versatile e will become clear.

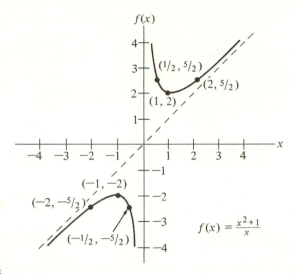

FIGURE 2.5

TABLE 2.14

Values of x	Values of f(x)	Values of x	Values of f(x)
1	2	5,001	2.71801 ...
2	2.25	9,801	2.71814 ...
3	2.37037 ...	20,001	2.71821 ...
10	2.59374 ...	35,001	2.71824 ...
25	2.66778 ...	50,001	2.71825 ...
100	2.70481 ...	65,001	2.71826 ...
501	2.71557 ...	80,001	2.71826 ...
991	2.71691 ...	100,001	2.71827 ...

It is also useful to examine the limit of a function as x increases without bound in the negative direction. Such a limit can be defined as follows:

DEFINITION

If the values of the dependent variable, $f(x)$, approach a single number L as x increases without bound in the negative direction, then L is the limit of the function as $x \to -\infty$. This is denoted by

$$\lim_{x \to -\infty} f(x) = L$$

Example 19   Let $f(x) = 1/x$. Find the limit of the function as $x \to -\infty$. Note that this is the same function as was used in Example 15. Therefore its graph is the same as Figure 2.3. From Table 2.15 we find that $\lim_{x \to -\infty} (1/x) = 0$.

TABLE 2.15

Values of x	Values of $f(x)$
-100	$-1/100 = -.01$
$-5,500$	$-1/5,500 = -.000181818\ldots$
$-150,000$	$-1/150,000 = -.00000666\ldots$
$-1,000,000$	$-1/1,000,000 = -.000001$

▶ **Exercise 2.4**

Find the limit of the function in each of the following problems, if it exists. Make your own table for each problem.

1 $\lim\limits_{x \to \infty} \dfrac{1}{x^2}$
    2 $\lim\limits_{x \to \infty} \dfrac{x+3}{x+4}$

3 $\lim\limits_{x \to \infty} \dfrac{3x-5}{2x+1}$
    4 $\lim\limits_{x \to \infty} \dfrac{x^2}{x+1}$

5 $\lim\limits_{x \to \infty} \dfrac{1}{\sqrt{x}}$
    6 $\lim\limits_{x \to \infty} \dfrac{x^2-4}{3x^2+2x}$

7 $\lim\limits_{x \to -\infty} \dfrac{x+1}{2x-1}$
    8 $\lim\limits_{x \to -\infty} \dfrac{x}{x+1}$

2.5   Techniques for Finding Limits as $x \to \pm\infty$

There are certain types of functions that occur often enough to warrant an investigation as to the possibility of finding the limit of the function without having to set up a table when $x \to \pm\infty$. In this section we will concentrate on rational functions and functions of the form $y = e^{ax}$, where a is some constant.

DEFINITION

A polynomial (of degree n) that is a function of one variable is of the following form:

$$P(x) = a_0 x^n + a_1 x^{n-1} + a_2 x^{n-2} + \cdots + a_{n-1}x + a_n$$

where n is a nonnegative integer and a_i is a real number for all i such that $0 \le i \le n$. ($a_0 \ne 0$.)

Example 20

| *Examples of polynomials* | | *Nonexamples of polynomials* |

$P(x) = x^3 + 3x^2 - 4$ (degree 3) $F(x) = 3\sqrt{x} + 4$

$f(x) = 3x - 5$ (degree 1) $g(x) = 3/x + 5x - 3$

$g(x) = 4x$ (degree 1) $H(x) = 4x^{-2} + 3x - 5$

$Q(x) = 8 + 3x - 6x^4 + 7x^5$ (degree 5)

DEFINITION

A rational function is the quotient of two polynomial functions.

Example 21   The following are examples of rational functions.

$$f(x) = \frac{x^3 - 3x^2 + 7}{2x - 7}$$

$$g(x) = \frac{x}{x^2 - 6}$$

$$h(x) = \frac{3x + 5}{2x - 3}$$

At this point we will investigate the existence of the limits of rational functions as $x \to \pm \infty$. The same technique will work whether $x \to \infty$ or $x \to -\infty$. If the limit exists, then this technique will also find the value of the limit. The method in this problem is to divide x^n into each term of both the dividend and the divisor, where n is the largest exponent of the variable in the rational function. Without restating them here, we would like to note that the six limit theorems stated earlier are also true as $x \to \infty$ or as $x \to -\infty$.

Example 22   Let $f(x) = (x + 1)/(2x - 1)$. Find the limit of the function as $x \to \infty$. Since the largest exponent is 1, we will divide each term by x^1, or simply x:

$$\lim_{x \to \infty} \frac{x + 1}{2x - 1} = \lim_{x \to \infty} \frac{x/x + 1/x}{2x/x - 1/x} = \lim_{x \to \infty} \frac{1 + 1/x}{2 - 1/x}$$

$$= \frac{\lim_{x \to \infty} 1 + \lim_{x \to \infty} (1/x)}{\lim_{x \to \infty} 2 - \lim_{x \to \infty} (1/x)} = \frac{1 + 0}{2 - 0} = \frac{1}{2}$$

[*Note:* $\lim_{x \to \infty} (1/x) = 0$ was established in Example 15.]

Example 23   Let $f(x) = (x^2 + 1)/x$. Find the limit of the function as $x \to \infty$. Dividing each term by x^2, we have

$$\lim_{x \to \infty} \frac{x^2 + 1}{x} = \lim_{x \to \infty} \frac{x^2/x^2 + 1/x^2}{x/x^2} = \lim_{x \to \infty} \frac{1 + 1/x^2}{1/x}$$

which appears to become $(1 + 0)/0$, which is undefined. In a rational function of this type, the undefined statement implies that the limit does not exist.

Example 24 Let $f(x) = x/(x^2 + 3)$. Find the limit of this function as $x \to \infty$. Dividing each term by x^2, we have

$$\lim_{x \to \infty} \frac{x}{x^2 + 3} = \lim_{x \to \infty} \frac{x/x^2}{x^2/x^2 + 3/x^2} = \lim_{x \to \infty} \frac{1/x}{1 + 3/x^2}$$

$$= \frac{\lim_{x \to \infty} (1/x)}{\lim_{x \to \infty} 1 + \lim_{x \to \infty} (3/x^2)} = \frac{0}{1 + 0} = 0$$

Summarizing the previous examples, three facts can be stated for limits of rational functions.

1 If the largest exponent of x in the numerator is equal to the largest exponent of x in the denominator, the limit is a/b, where a and b are the numerical coefficients of the terms involving this largest exponent in the numerator and denominator, respectively.

Example 25 Let $f(x) = (4x^3 - 6x^2 + 5x - 2)/(3x^3 + 5x + 1)$. Find the limit of the function as $x \to \infty$. Since the largest exponent in the numerator is 3 and the largest exponent in the denominator is 3, we look at the numerical coefficients of x^3 in both the numerator and the denominator. Thus, $\lim_{x \to \infty} f(x) = 4/3$.

2 If the largest exponent in the numerator is greater than the largest exponent in the denominator, then the limit does not exist.

Example 26 Let $f(x) = (x^4 + 3x)/(x^2 + 5x - 6)$. Find the limit of the function as $x \to \infty$. Since the exponent 4 in the numerator is larger than the exponent 2 in the denominator, $\lim_{x \to \infty} f(x)$ does not exist.

3 If the largest exponent in the denominator is greater than the largest exponent in the numerator, then the limit of the function is zero.

Example 27 Let $f(x) = (x + 2)/x^3$. Find the limit of the function as $x \to \infty$. Since the exponent 3 in the denominator is greater than the exponent 1 in the numerator, $\lim_{x \to \infty} f(x) = 0$.

Now we want to investigate the limit of a function, called the *exponential function*, of the type $f(x) = e^{ax}$, where a is some constant and x is approaching $\pm \infty$.

Example 28 Let $f(x) = e^{.3x}$. Find the limit of the function as $x \to \infty$ and as $x \to -\infty$. First, we will establish a table of values for x and the corresponding values of $f(x)$. (See Table 2.16.) In order to find the values of $f(x)$ in this table, we may use a table of powers for e, or some type of electronic calculator.

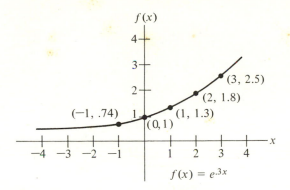

$$f(x) = e^{.3x}$$

FIGURE 2.6

TABLE 2.16

Values of x	Values of f(x)	Values of x	Values of f(x)
1	1.34986	−1	.74082
10	20.0855	−10	.04979
100	10,686,000,000,000.	−50	.0000003059
500	1.3937×10^{65}	−100	.00000000000093576

Thus $\lim_{x \to \infty} e^{.3x}$ does not exist, but $\lim_{x \to -\infty} e^{.3x} = 0$. (See Figure 2.6.)

Example 29 Let $f(x) = e^{-.05x}$. Find the limit of the function as $x \to \infty$ and as $x \to -\infty$.

TABLE 2.17

Values of x	Values of f(x)	Values of x	Values of f(x)
10	.6065	−10	1.6487
50	.0821	−50	12.1825
100	.006738	−100	148.413
1000	1.929×10^{-22}	−1000	5.185×10^{21}

Therefore $\lim_{x \to \infty} e^{-.05x} = 0$, but $\lim_{x \to -\infty} e^{-.05x}$ does not exist. (See Table 2.17 and Figure 2.7.)

Tables 2.16 and 2.17 can be generalized as follows:

1 If $f(x) = e^{ax}$ and a is positive, then $\lim_{x \to \infty} e^{ax}$ does not exist and $\lim_{x \to -\infty} e^{ax} = 0$.

2 If $f(x) = e^{ax}$ and a is negative, then $\lim_{x \to \infty} e^{ax} = 0$ and $\lim_{x \to -\infty} e^{ax}$ does not exist.

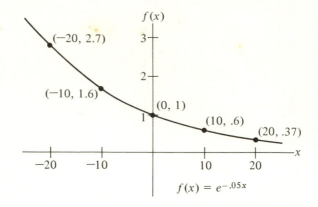

$$f(x) = e^{-.05x}$$

FIGURE 2.7

▶ **Exercise 2.5**

Find the limits of each of the following functions, if possible.

1 $\lim\limits_{x \to \infty} \dfrac{2x^2 + 3x + 1}{3x^2 - x - 5}$

2 $\lim\limits_{x \to \infty} \dfrac{x^2 + 1}{2x^2 + 6x - 8}$

3 $\lim\limits_{x \to \infty} \dfrac{3x}{4x^2 + 2x - 8}$

4 $\lim\limits_{x \to \infty} \dfrac{x^3 + 4x^2 + 5x - 3}{3x^2 - 5x + 6}$

5 $\lim\limits_{x \to \infty} \dfrac{3x + 2}{2x - 5}$

6 $\lim\limits_{x \to \infty} e^{-.4x}$

7 $\lim\limits_{x \to \infty} (3 + e^{-.02x})$

8 $\lim\limits_{x \to \infty} (1200 - 1200e^{-.4x})$

9 $\lim\limits_{x \to \infty} \dfrac{x}{x^2 + 1}$

10 $\lim\limits_{x \to -\infty} \dfrac{3x + 2}{2x - 5}$

11 $\lim\limits_{x \to -\infty} \dfrac{1}{x^2}$

12 $\lim\limits_{x \to -\infty} (1 + e^{2x})$

2.6 Application of Limits

The concept of a limit has some practical applications. Two such applications deal with the learning curve and with continuously compounded interest.

An equation of the type $y = a - ae^{-bx}$, where a and b are positive constants, is commonly called a learning curve. It derives its name from the fact that learning increases rapidly at first but gradually increases at a slower and slower rate and finally approaches an upper limit. This concept may also apply to certain business problems, as illustrated in Applied Example 30.

Applied Example 30   The revenue function associated with a certain product is given by $R(x) = 15{,}000 - 15{,}000e^{-.03x}$, where x is the amount of advertising dollars in hundreds of dollars.

1. Find the revenue when $x = 10$.
$$R(10) = 15{,}000 - 15{,}000e^{-.03(10)} = 15{,}000 - 11{,}115 = 3885$$

2. Find the revenue when $x = 50$.
$$R(50) = 15{,}000 - 15{,}000e^{-.03(50)} = 15{,}000 - 3347 = 11{,}653$$

3. Find the revenue when $x = 100$.
$$R(100) = 15{,}000 - 15{,}000e^{-.03(100)} = 15{,}000 - 737 = 14{,}263$$

4. What is the maximum expected revenue?
$$\lim_{x \to \infty} R(x) = \lim_{x \to \infty} (15{,}000 - 15{,}000e^{-.03x})$$
$$= \lim_{x \to \infty} 15{,}000 - \lim_{x \to \infty} 15{,}000e^{-.03x}$$
$$= \lim_{x \to \infty} 15{,}000 - \left(\lim_{x \to \infty} 15{,}000\right)\left(\lim_{x \to \infty} e^{-.03x}\right)$$
$$= 15{,}000 - 15{,}000(0) = 15{,}000.$$

5. What percentage of the maximum expected revenue is obtained at $x = 100$?

At $x = 100$, $R(100) = 14{,}263$. Since the maximum expected revenue is 15,000, the percentage of this revenue that has been reached at $x = 100$ is $14{,}263/15{,}000 = .951 = 95.1\%$. This means that very little increase in revenue can be expected for increases in advertising beyond $x = 100$.

6. Graph $R(x) = 15{,}000 - 15{,}000e^{-.03x}$. (See Figure 2.8.)

FIGURE 2.8

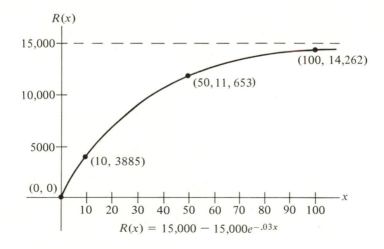

$$R(x) = 15{,}000 - 15{,}000e^{-.03x}$$

Applied Example 31   The compound interest formula yields an amount A for principal P invested for t years as given by the equation, $A = P(1 + r/x)^{xt}$, where r is the rate of interest per year, x is the number of times the interest is compounded per year, and t is the number of years. As the number of compounding periods x increases, the amount A increases. However, A does not increase without bound. Since $A = P(1 + r/x)^{xt}$, we want to find the value of A as the number of compounding periods x approaches infinity. Thus $\lim_{x \to \infty} A = \lim_{x \to \infty} P(1 + r/x)^{xt} = P \lim_{x \to \infty} (1 + r/x)^{xt}$. In order to evaluate $\lim_{x \to \infty} (1 + r/x)^{xt}$, let $N = x/r$. If $N = x/r$, the following three statements are true:

$$1 \quad x = Nr$$
$$2 \quad 1/N = r/x$$
$$3 \quad \text{As } x \to \infty, N \to \infty$$

Thus, by substitution we have

$$\lim_{x \to \infty} \left(1 + \frac{r}{x}\right)^{xt} = \lim_{N \to \infty} \left(1 + \frac{1}{N}\right)^{Nrt} = \left(\lim_{N \to \infty} \left(1 + \frac{1}{N}\right)^{N}\right)^{rt} = e^{rt}$$

Therefore

$$\lim_{x \to \infty} A = P \lim_{x \to \infty} \left(1 + \frac{r}{x}\right)^{xt} = P(e)^{rt}$$

Applied Example 32   What will be the amount received if $1000 is invested at 6% compounded continuously for 1 year?

$$A = 1000e^{.06} \doteq 1000(1.06184) = \$1061.84$$

If the interest were compounded annually for 1 year, then A would equal $1060.00. Thus, increasing the number of compounding periods from one to infinity yields an extra $1.84 per thousand dollars under the given conditions.

▶ **Exercise 2.6**

1 A man deposits $1500 in an account that pays 6% annually. If the interest is compounded continuously, how much will he have in the account after 5 years?

2 Mr. Smith wants to have savings amounting to $6000 10 years from now. If this savings account pays 5% compounded continuously, how much should he deposit now in order to have the $6000 in 10 years?

3 On a particular scale a certain man's learning curve can be expressed as $f(t) = 12 - 12e^{-.15t}$, where t is in years.
  a Find the optimum value the man approaches on this scale.
  b Find the value on the scale that he has reached after 10 years.
  c What percentage of the optimum value has been reached after 10 years?

4 The revenue function for a certain product is given by $R(x) = 10,000 - 10,000e^{-.05x}$, where x is the amount, in hundreds of dollars, spent for promotion of the product.

a Find the revenue at $x = 50$.

b Find the maximum revenue that can be expected if the promotional expense is unlimited.

c What percentage of the maximum revenue is obtained at $x = 50$?

5 Let $R(x)$ be the revenue, in millions of dollars, yielded from the sale of product Y, and let x be the number of units produced. Market research has shown that the revenue is related to the number of units produced by the following equation: $R(x) = 3x/(x + 1)$. What is the maximum revenue that can be expected as production increases without bound?

6 A plot of land is at present worth \$500. If the value V increases with the passage of time t according to the equation $V = 500[1 + 2t/(t + 5)]$, what is the maximum value the land will approach?

7 A man assembled a jigsaw puzzle in 5 hr on his first attempt. Each successive time he assembled the puzzle he decreased his time t according to $t = 2.5(1 + e^{-.5x+.5})$, where x is the number of times he has assembled the puzzle. What is the minimum amount of time in which he can expect to assemble the puzzle? How long did it take him to assemble the puzzle the second time?

8 The number of items y manufactured per day, x days after the start of production, is $y = 40(1 - e^{-.2x})$. What is the maximum number of units that can be produced per day? After 10 days' production, what percentage of the maximum is actually being produced?

9 With present facilities, the average cost per unit y of producing x ball-point pens is $y = (x^2 + 2x - 3)/(x^2 - x - 2)$, where $x > 3$. Find the minimum cost per unit the company may expect if production is increased without bound.

10 The capacitance of an electric circuit is the quantity of electricity that can be stored in the circuit. For a circuit containing a capacitance in series with a resistance and under a constant electromotive force, the quantity of electricity stored in the circuit increases from zero to full capacity as given by the equation $q = EC(1 - e^{-t/RC})$ where E is the electromotive force, C is the capacitance of the circuit, R is the resistance, q is the quantity of electricity, and t is the time elapsed after the switch is closed. For a given circuit all letters represent positive constants except t and q. What is the maximum value approached by q?

11 The population of a town is at present 10,000. If the town can expect a continuously compounded growth of 1% per year, what will be the population of the town in 10 years?

2.7 Continuity

In Exercise 2.3 the evaluation of some of the limits required factoring and simplifying while others did not. That is, some of the functions $y = f(x)$ had limits equal to $f(p)$ as x approached p, while others did not.

Example 33 Let $f(x) = 2x + 5$. The limit of $2x + 5$ is 7 as x approaches 1 and $f(1) = 7$. Therefore $\lim_{x \to 1} (2x + 5) = f(1)$.

Example 34 Let $f(n) = 4n^2 + 2$. Then $\lim_{n \to 2} (4n^2 + 2) = f(2) = 18$.

Example 35 Let $f(x) = (x^2 - 5x + 6)/(x - 2)$. In this function $f(2) = 0/0$, which is meaningless, while

$$\lim_{x \to 2} \frac{x^2 - 5x + 6}{x - 2} = -1$$

Therefore the limit of the function as x approaches 2 is not equal to $f(2)$.

What property of functions links together some functional values and some limits of functions? The property is called *continuity*.

DEFINITION

The function $y = f(x)$ is continuous at a point p if the following three conditions are true:

1 $f(p)$ is defined and finite.

2 $\lim_{x \to p} f(x)$ exists and is finite.

3 $\lim_{x \to p} f(x) = f(p)$.

If *any one* of the conditions fails to be true, then the function is not continuous (is discontinuous) at the point p.

We want to consider the following situations when dealing with functions, limits, and continuity.

First, $f(p)$ may be defined and $\lim_{x \to p} f(x)$ does not exist.

Example 36 Show that

$$f(x) = \begin{cases} x^2 & \text{if } x \le 0 \\ 2x + 1 & \text{if } x > 0 \end{cases}$$

is discontinuous at $x = 0$.

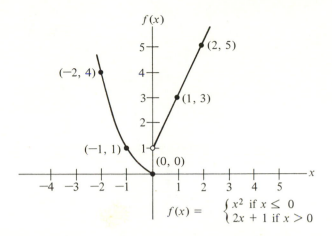

$$f(x) = \begin{cases} x^2 & \text{if } x \leq 0 \\ 2x + 1 & \text{if } x > 0 \end{cases}$$

FIGURE 2.9

1. $f(0) = 0$
2. $\lim_{x \to 0} f(x)$ does not exist

The function has a right-hand limit of 1 and a left-hand limit of zero as x approaches zero. Therefore the function has no limit, because the right-hand limit and left-hand limit are not equal. (See Figure 2.9.)

Secondly, $f(p)$ may be defined and the limit may exist, but $f(p) \neq \lim_{x \to p} f(x)$.

Example 37 Show that

$$f(x) = \begin{cases} x^3 & \text{if } x \neq 0 \\ 2 & \text{if } x = 0 \end{cases}$$

is discontinuous at $x = 0$.

1. $f(0) = 2$
2. $\lim_{x \to 0} f(x) = 0$
3. $2 \neq 0$.

Therefore the function is discontinuous at $x = 0$. (See Figure 2.10.)

Thirdly, $\lim_{x \to p} f(x)$ may exist and $f(p)$ not be defined.

Example 38 Show that $f(x) = (x^2 - x - 6)/(x - 3)$ is discontinuous at $x = 3$. We have $f(3) = 0/0$ is meaningless. Therefore the function is discontinuous at $x = 3$. However, $\lim_{x \to 3} f(x) = 5$.

Geometrically, if a function is discontinuous at a point, the graph is "broken" at that point.

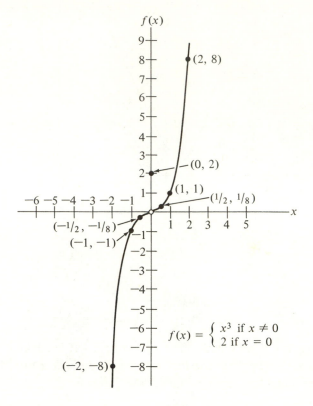

$$f(x) = \begin{cases} x^3 & \text{if } x \neq 0 \\ 2 & \text{if } x = 0 \end{cases}$$

FIGURE 2.10

In each of the previous cases functional values and/or limits of functions exist but continuity does not. Finally, if all three statements in the definition of a continuous function are true, we have a fusing together of limits of functions and functional evaluations in a "well-behaved" class of functions called *continuous functions*. (See Figure 2.11.)

Example 39 Show that

$$f(x) = \begin{cases} 3x & \text{if } x > 1 \\ 2x^2 + 1 & \text{if } x \leq 1 \end{cases}$$

is continuous at $x = 1$. (See Figure 2.12.)

1. $f(1) = 3$
2. $\lim_{x \to 1} f(x) = 3$
3. $\lim_{x \to 1} f(x) = f(1)$ since $3 = 3$

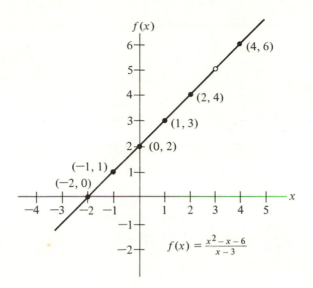

FIGURE 2.11

Example 40   Let $f(x) = x^2$. Show that $f(x) = x^2$ is continuous at $x = 0$.

   1. $f(0) = 0$
   2. $\lim_{x \to 0} f(x) = 0$
   3. $\lim_{x \to 0} f(x) = f(0)$, since $0 = 0$

FIGURE 2.12

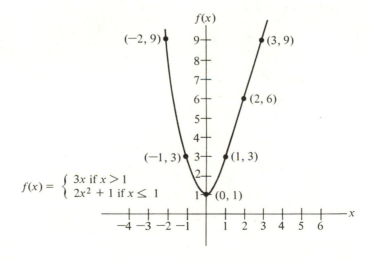

Note: All polynomials with unrestricted domain are continuous at all points of their domain.

▶ **Exercise 2.7**

1 Show that $y = 2x^2 - 3$ is continuous at $x = 1$.

2 Show that $y = 1/(1 - x)$ is continuous at $x = 2$.

3 Show that $y = 1/(1 - x)$ is discontinuous at $x = 1$.

4 Show that

$$f(x) = \begin{cases} x & \text{if } x \neq 0 \\ 3 & \text{if } x = 0 \end{cases}$$

is discontinuous at $x = 0$.

5 Show that $g(x) = (x^2 + 3x - 10)/(x - 2)$ is discontinuous at $x = 2$.

6 Show that

$$h(x) = \begin{cases} 2x & \text{if } x < 2 \\ x^2 & \text{if } x \geq 2 \end{cases}$$

is continuous at $x = 2$.

7 Show that

$$y = \begin{cases} x + 3 & \text{if } x < 2 \\ x^2 & \text{if } x \geq 2 \end{cases}$$

is discontinuous at $x = 2$.

8 Show that

$$h(n) = \begin{cases} n + 1 & \text{if } n > 1 \\ 3 & \text{if } n = 1 \\ 3n + 1 & \text{if } n < 1 \end{cases}$$

is discontinuous at $n = 1$.

9 Show that

$$y = \begin{cases} 2^x & \text{if } x > 0 \\ 0 & \text{if } x = 0 \\ 1 - x & \text{if } x < 0 \end{cases}$$

is discontinuous at $x = 0$.

10 At what value of x is $g(x) = 2/(x - 3)$ discontinuous? Why?

11 At what value of x is $y = (x^2 - 3x - 4)/(x - 4)$ discontinuous? Why?

12 At what value of x (if any) is $y = -3x + 4$ discontinuous? Why?

---

self-test · chapter two

Find the limit in Problems 1 through 6.

1 $\lim\limits_{x \to 2} (3x + 2)$

2 $\lim\limits_{x \to 3} \dfrac{x^2 - 8x + 15}{x - 3}$

3 $\lim\limits_{x \to \infty} \dfrac{2x + 3}{3x - 1}$

4 $\lim\limits_{x \to \infty} \dfrac{1}{x^3}$

5 $\lim\limits_{x \to \infty} (140 - 140e^{-.2x})$

6 $\lim\limits_{x \to 2} \dfrac{3}{x - 2}$

7 Let $g(x) = \begin{cases} 2x + 1 & \text{if } x \geq 6 \\ x - 1 & \text{if } x < 6 \end{cases}$

  a Find the right-hand limit as $x \to 6$.
  b Find the left-hand limit of the function as $x \to 6$.
  c Find the limit of the function as $x \to 6$, if possible.

8 Find the values of x for which $f(x) = (x^2 + 3x + 5)/(x - 5)$ is discontinuous.

9 Is

$$f(x) = \begin{cases} x^2 & \text{if } x \geq 1 \\ 2x - 1 & \text{if } x < 1 \end{cases}$$

continuous at $x = 1$? Explain your answer.

chapter three

the derivative with applications 1

3.1 The Concept of the Derivative

In the first two chapters we discussed functions, slopes, limits of functions, and continuous functions. Now we will attempt to blend these topics together to produce a more powerful weapon in the arsenal of mathematics.

Let us review the concept of slope. If two points on the graph of a function are connected with a straight line, the slope of the line can be determined from the coordinates of the points. We called the slope the average rate of change of the function with respect to a change in the independent variable.

However, this average rate of change of the function relating two points does not yield much information about the rate of change of the function at a single point. The desirability of being able to determine the rate of change of the function at a single point leads to the concept of the derivative.

In Example 1 of this chapter we take a succession of points on the graph of $y = x^2/4$ that approach the point $(1, \frac{1}{4})$. Then we find the slopes of the lines (called *secant lines*) through the point $(1, \frac{1}{4})$ and each successive point. From the slopes (of secant lines), by the use of limits, we can determine the slope of the (tangent) line through the single point $(1, \frac{1}{4})$. Study Example 1 carefully.

Example 1 Let $f(x) = x^2/4$. Let $(p, f(p))$ be the fixed point $(1, \frac{1}{4})$. Find the slope of the secant lines through the point $(1, \frac{1}{4})$ and the other points on the curve that approach $(1, \frac{1}{4})$. The slope of the secant lines is denoted m_{sec}. (See Figure 3.1 and Table 3.1.)

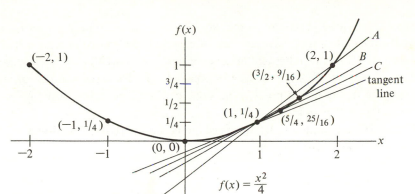

FIGURE 3.1

TABLE 3.1

Fixed point $(p, f(p))$	Point approaching fixed point	$m_{sec} = \dfrac{f(x) - f(p)}{x - p}$	
(1, 1/4)	(2, 1)	.7500	Line A
(1, 1/4)	(3/2, 9/16)	.6250	Line B
(1, 1/4)	(5/4, 25/64)	.5625	Line C
(1, 1/4)	(101/100, 10,201/40,000)	.5025	

As $x \to 1$ from the right, $m_{sec} \to .5000$

(1, 1/4)	(1/2, 1/16)	.3750	
(1, 1/4)	(4/5, 4/25)	.4500	
(1, 1/4)	(9/10, 81/400)	.4750	
(1, 1/4)	(99/100, 9,801/40,000)	.4975	

As $x \to 1$ from the left, $m_{sec} \to .5000$

Therefore,

$$\lim_{x \to 1} \frac{f(x) - f(1)}{x - 1} = \frac{1}{2}$$

What is this $\frac{1}{2}$? First it is the limit of the successive slopes of the secant lines. Thus it seems logical that $\frac{1}{2}$ is a slope; but the slope of what?

As the points in Example 1 get closer to the fixed point $(1, \frac{1}{4})$, the secant lines approach a limiting line. This limiting line is called the *tangent line* to $f(x) = x^2/4$ at the point $(1, \frac{1}{4})$. The value $\frac{1}{2}$, which is the limit of the slopes of the secant lines, is the slope of the tangent line to $f(x) = x^2/4$ through the point $(1, \frac{1}{4})$.

Since $\frac{1}{2}$ is a slope, it represents the rate of change of the function with respect to x. However, it is not an average rate of change between two points on $f(x) = x^2/4$, since $(1, \frac{1}{4})$ is the only point under consideration. We call $\frac{1}{2}$ the *instantaneous rate of change* of $f(x) = x^2/4$ at the point $(1, \frac{1}{4})$. The *instantaneous rate of change* of a function at a point is the slope of the tangent line to the graph of the function at that point.

Generalizing Example 1, let the fixed point be $(p, f(p))$. The slopes of the secant lines can be expressed as

$$m_{\text{sec}} = \frac{f(x) - f(p)}{x - p}$$

where $(x, f(x))$ is any point on the graph of the function except $(p, f(p))$. Note that m_{sec} is a function of x. Taking the limit as x approaches p, we have the slope of the tangent line (provided the limit exists), denoted

$$m_{\text{tan}} = \lim_{x \to p} m_{\text{sec}} = \lim_{x \to p} \frac{f(x) - f(p)}{x - p}$$

Now the derivative can be defined.

DEFINITION

The derivative, denoted $f'(p)$, of the function $y = f(x)$ at the point $(p, f(p))$ is

$$f'(p) = \lim_{x \to p} \frac{f(x) - f(p)}{x - p}$$

provided the limit exists.

Geometrically, the derivative evaluated at a point on a curve gives the slope of the tangent line to the curve at that point.

The following examples use the definition of the derivative to find the derivative of a given function at $x = p$.

Example 2 Let $f(x) = 4$. Find the derivative at p.

$$f'(p) = \lim_{x \to p} \frac{f(x) - f(p)}{x - p} = \lim_{x \to p} \frac{4 - 4}{x - p} = \lim_{x \to p} \frac{0}{x - p} = \lim_{x \to p} 0 = 0$$

Since $f(x) = 4$ is a constant function, its graph has a slope of zero at all points.

Example 3 Let $f(x) = x^2$. Find $f'(p)$.

$$f'(p) = \lim_{x \to p} \frac{f(x) - f(p)}{x - p} = \lim_{x \to p} \frac{x^2 - p^2}{x - p} = \lim_{x \to p} \frac{(x + p)(x - p)}{x - p}$$
$$= \lim_{x \to p} (x + p) = 2p$$

If a specific value of p is given, for example $p = 2$, then $f'(p) = 2p$ can be written $f'(2) = 2(2) = 4$. Then 4 is the slope of the tangent line to the graph of $f(x) = x^2$ at the point $(2, 4)$. (See Figure 3.2.) If no particular value of p is given, then p is usually replaced by x. This yields the general form of the derivative $f'(x) = 2x$, which is sometimes called the *derived function*.

Example 4 Let $f(x) = 4x^2 - 2x + 3$. Find $f'(x)$.

$$f'(p) = \lim_{x \to p} \frac{f(x) - f(p)}{x - p}$$

$$= \lim_{x \to p} \frac{(4x^2 - 2x + 3) - (4p^2 - 2p + 3)}{x - p}$$

$$= \lim_{x \to p} \frac{4x^2 - 2x + 3 - 4p^2 + 2p - 3}{x - p}$$

$$= \lim_{x \to p} \frac{4x^2 - 4p^2 - 2x + 2p}{x - p}$$

$$= \lim_{x \to p} \frac{4(x^2 - p^2) - 2(x - p)}{x - p}$$

$$= \lim_{x \to p} \frac{4(x + p)(x - p) - 2(x - p)}{x - p}$$

$$= \lim_{x \to p} \frac{(x - p)[4(x + p) - 2]}{x - p}$$

$$= \lim_{x \to p} [4(x + p) - 2]$$

$$= 4(p + p) - 2 = 8p - 2$$

Therefore $f'(x) = 8x - 2$.

FIGURE 3.2

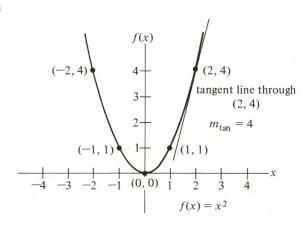

$f(x) = x^2$

Other notations used to represent the derivative are y' and dy/dx. That is,

$$y' = \frac{dy}{dx} = f'(x) = \lim_{x \to p} \frac{f(x) - f(p)}{x - p}$$

▶ **Exercise 3.1**

1 Let $f(x) = 2x^2 + 1$. Let $(1, 3)$ be the fixed point.
   a Complete Table 3.2.
   b Find $f'(x)$ for $f(x) = 2x^2 + 1$ by using the definition of the derivative.

TABLE 3.2

Fixed point	Other points	Slope of secant line
(1, 3)	(1.5,)	_____
(1, 3)	(1.3,)	_____
(1, 3)	(1.1,)	_____
(1, 3)	(1.01,)	_____
	As $x \to 1$ from the right, $m_{\text{sec}} \to$	_____
(1, 3)	(.5,)	_____
(1, 3)	(.8,)	_____
(1, 3)	(.9,)	_____
(1, 3)	(.99,)	_____
	As $x \to 1$ from the left, $m_{\text{sec}} \to$	_____

   c Evaluate the derivative at $(1, 3)$.
   d Does the value determined in (c) agree with the limit observed in (a)?
   e What is the slope of the tangent line to the curve at the point $(1, 3)$?

2 Find the derivative for each of the following by using the definition.
   a $y = x$ **b** $g(x) = 3x + 4$
   c $f(x) = 3x^2 + 2$ **d** $f(x) = 2x^2 + 3x - 3$
   e $f(x) = 1/x$

3 For each part of Problem 2, write the equation of the tangent line at the point where $x = 2$.

4 For each part of Problem 2, find the value(s) of x where the slope of the tangent line is 1.

3.2   Rules for Finding Derivatives (Differentiation)

After having used the definition to find the derivative, one might ask, Is there an easier method that can be used to find the derivative? Yes. Functions can be considered to exist in classes. Some of these classes are polynomials, rational

functions, powers of polynomials, exponential functions, and so on. In finding
the derivative there are certain patterns that develop for each class. These
patterns lead to rules for differentiation. In this chapter, four rules are stated.
Four more rules will be added in Chapter 4. Each rule is stated symbolically,
then restated without symbols. The rules for finding the derivative of each class
will be established by using the definition. From now on, our interest in the
definition will be restricted to establishing rules for differentiating an entire
class of functions.

RULE 1

Let $f(x) = c$, where c is a constant. Then $f'(x) = 0$.

   The derivative of a constant function is zero.

Proof   By definition,

$$f'(p) = \lim_{x \to p} \frac{f(x) - f(p)}{x - p} = \lim_{x \to p} \frac{c - c}{x - p}$$

$$= \lim_{x \to p} \frac{0}{x - p} = \lim_{x \to p} 0 = 0$$

Example 5

   If $y = -3$, then $y' = 0$.

   If $f(x) = 5$, then $f'(x) = 0$.

   If $y = 0$, then $dy/dx = 0$.

   If $g(x) = 2^3$, then $g'(x) = 0$.

RULE 2

If $f(x) = x^n$, then $f'(x) = nx^{n-1}$ for n any real number.

   The derivative of x raised to an exponent is the exponent multiplied by x to
the exponent minus 1.

   Before proving Rule 2, where n is a counting number, it is helpful to review
some special factoring patterns. Note the following pattern:

$$x^2 - y^2 = (x - y)\underbrace{(x + y)}_{\text{two terms}}$$

$$x^3 - y^3 = (x - y)(x^2 + xy + y^2)$$

$$\underbrace{}_{\text{three terms}}$$

$$x^4 - y^4 = (x - y)(x^3 + x^2y + xy^2 + y^3)$$

$$\underbrace{}_{\text{four terms}}$$

Following the above pattern, we have

$$x^n - y^n = (x - y)\underbrace{(x^{n-1} + x^{n-2}y + x^{n-3}y^2 + \cdots + y^{n-1})}_{n \text{ terms}}$$

where n is any counting number. Now the proof of Rule 2 can be given, where n is restricted to a counting number.

Proof

$$f'(p) = \lim_{x \to p} \frac{f(x) - f(p)}{x - p}$$

$$= \lim_{x \to p} \frac{x^n - p^n}{x - p}$$

$$= \lim_{x \to p} \frac{(x - p)(x^{n-1} + x^{n-2}p + x^{n-3}p^2 + \cdots + p^{n-1})}{x - p}$$

$$= \underbrace{p^{n-1} + p^{n-2}p + p^{n-3}p^2 + \cdots + p^{n-1}}_{n \text{ terms}}$$

$$= np^{n-1}$$

Using x for p, $f'(x) = nx^{n-1}$. Though the proof is limited to counting numbers, Rule 2 is true for any constant exponent.

Example 6

> If $y = x^5$, then $y' = 5x^4$.
>
> If $f(x) = x^{1/2}$, then $f'(x) = \frac{1}{2}x^{-1/2}$.
>
> If $y = 1/x^2 = x^{-2}$, then $dy/dx = -2x^{-3} = -2/x^3$.

(A review of exponents is given in Appendix 1.)

RULE 3

If $f(x) = c \cdot g(x)$, where c is a constant, then $f'(x) = c \cdot g'(x)$, where $g(x)$ has a derivative.

The derivative of a constant times a differentiable expression is the constant multiplied by the derivative of the expression.

Proof Let $f(x) = c \cdot g(x)$, where c is a constant. Then

$$f'(p) = \lim_{x \to p} \frac{c \cdot g(x) - c \cdot g(p)}{x - p}$$

By Theorem 2 of limits,

$$f'(p) = c \cdot \lim_{x \to p} \frac{g(x) - g(p)}{x - p}$$

By the definition of the derivative,

$$f'(p) = c \cdot g'(p)$$

Replacing p by x,

$$f'(x) = c \cdot g'(x)$$

Example 7

If $y = 3x^5$, then $y' = 15x^4$.

If $f(x) = 2x^{1/2}$, then $f'(x) = x^{-1/2} = 1/\sqrt{x}$.

RULE 4

If $y = f(x) + g(x) + \cdots + j(x)$, then $dy/dx = f'(x) + g'(x) + \cdots + j'(x)$, provided the derivative of each term exists.

If a function consists of a finite number of differentiable terms, the derivative is found by computing the derivative of each term separately and then adding the derivatives.

Proof Let $y = f(x) + g(x) + \cdots + j(x)$. Then

$$y' = \lim_{x \to p} \frac{f(x) + g(x) + \cdots + j(x) - [f(p) + g(p) + \cdots + j(p)]}{x - p}$$

Simplifying and applying Theorem 3 of limits, we have

$$y' = \lim_{x \to p} \frac{f(x) - f(p)}{x - p} + \lim_{x \to p} \frac{g(x) - g(p)}{x - p} + \cdots + \lim_{x \to p} \frac{j(x) - j(p)}{x - p}$$

By the definition of the derivative,

$$y' = f'(p) + g'(p) + \cdots + j'(p)$$

Replacing p by x,

$$y' = f'(x) + g'(x) + \cdots + j'(x)$$

Example 8

If $y = 3x^2 + 2x - 5$, then $y' = 6x + 2$.

If $p = 4q^2 - 6q$, then $p' = 8q - 6$.

If $c(x) = x^3 - 2x$, then $c'(x) = 3x^2 - 2$.

If $f(x) = 2/x - 3\sqrt{x} - 8$, then $f'(x) = -2/x^2 - 3/(2\sqrt{x})$.

▶ **Exercise 3.2**

Differentiate each of the functions in Problems 1 through 15.

1 $f(x) = 2x^3 - 6x^2 + x$

2 $y = 200 - x^2 + 4x^3$

3 $y = 2x^{1/4}$

4 $y = \dfrac{1}{x^2} - 6x$

5 $y = \dfrac{(x^2 + 6x)}{5}$

6 $y = \frac{1}{2}x^3 + \sqrt{x} + 5$

7 $y = \sqrt[3]{x}$

8 $y = 2\sqrt{x^3} + \dfrac{5}{x^2}$

9 $f(x) = 4x^3 - 6x^2 + 7x - 3$

10 $f(x) = \dfrac{x^3 + 2}{x}$ (*Hint:* Write as a sum.)

11 $f(x) = 4x^5 - 5x^4 + 6x^3 - 7x^2 + 6x + 3$

12 $g(x) = 14$

13 $g(x) = x$

14 $g(x) = 13x^2 + 4x - \dfrac{1}{x}$

15 $g(x) = 6\sqrt{x}$

16 If $f(x) = 2x^3 - 5x^2 + 6$, find $f'(2)$. What does the value represent?

17 If $f(x) = 6x^4 - 8x^3 + 3x - 7$, find $f(0)$ and $f'(0)$.

18 If $f(x) = x^2 - 6x + 8$, find the point on the graph where $f'(x) = 0$.

19 If

$$f(x) = \frac{x^3}{3} - 3x^2 + 8x - 1,$$

find the points on the graph of the function where $f'(x) = 0$.

20 If $y = x^2 + 3x - 4$, find the point on the graph where $f'(x) = 7$. Write the equation of the tangent line to the curve through this point.

3.3 Applications of the Derivative

In this section the interpretation of the derivative as a rate of change of the function with respect to the independent variable is most useful. There are several important definitions given in the following examples, and the reader should study them carefully.

Applied Example 9 A company has determined that its profit P is related to the number n of units produced by the equation $P = 100n - n^2$. Note that the domain of the function is restricted to nonnegative whole numbers by the application. However, in order to use calculus techniques, we will consider the function to be defined for all nonnegative real numbers.

1 Find the marginal profit function.

The marginal profit is the rate of change of profit with respect to production. Since the derivative gives the rate of change of the function, we define the marginal profit to be the derivative of the profit with respect to production. Therefore the marginal profit is $P' = 100 - 2n$.

2 What is the marginal profit at $n = 20$?
At $n = 20$, $P' = 100 - 2(20) = 60$.

3 At what level of production will the rate of change of profit be zero?
If $P' = 100 - 2n = 0$, then $n = 50$.

4 What is the rate of change of profit at $n = 80$?
At $n = 80$, $P' = 100 - 2(80) = -60$.

5 Graph $P = 100n - n^2$. (See Figure 3.3.)

Applied Example 10 The cost c per unit, called the average cost, is related to the number of units produced, x, by the equation $c = .01x^2 - .5x + 81.25$.

FIGURE 3.3

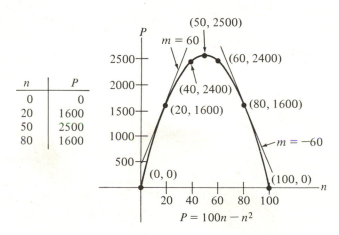

n	P
0	0
20	1600
50	2500
80	1600

$P = 100n - n^2$

1. Find the rate of change of cost per unit with respect to production.

$$c' = .02x - .5$$

2. At what level of production is the rate of change of cost per unit equal to zero?

If $c' = .02x - .5 = 0$, then $x = 25$.

3. What is the rate of change of cost per unit at $x = 15$?

$$c' = .02(15) - .5 = -.2$$

4. What is the rate of change of the cost per unit at $x = 35$?

$$c' = .02(35) - .5 = .2$$

5. Graph $c = .01x^2 - .5x + 81.25$. (See Figure 3.4.) The minimum cost per unit is 75.

Applied Example 11   A tank is filled by a small inlet pipe. The volume v in gallons in the tank at any time t in hours is $v = 300t^2 + 300t$.

1. What is the volume in the tank at the end of 1 hr?

At $t = 1$, $v = 300(1)^2 + 300(1) = 600$ gal

2. What is the volume in the tank at the end of 3 hr?

At $t = 3$, $v = 300(3)^2 + 300(3) = 3600$ gal

3. What is the average rate of change in volume as the time changes from 1 to 3 hr?

$$\frac{v(3) - v(1)}{3 - 1} = \frac{3600 - 600}{2} = 1500 \text{ gal/hr}$$

FIGURE 3.4

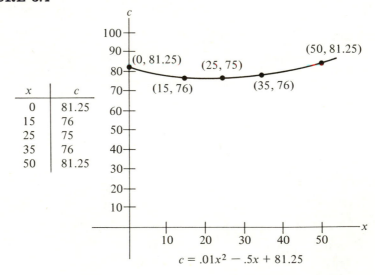

x	c
0	81.25
15	76
25	75
35	76
50	81.25

$c = .01x^2 - .5x + 81.25$

4. What is the instantaneous rate of change of volume at $t = 1$?

Since the derivative is the instantaneous rate of change, we must find dv/dt. Now $dv/dt = 600t + 300$. Therefore at $t = 1$, $dv/dt = 600(1) + 300 = 900$ gal/hr.

Applied Example 12 Suppose the distance S in feet that an object is displaced in time t in seconds is given by $S = 2t^2 - t$. Find the expression for the rate of change of distance with respect to time. That is, find dS/dt.

$$\frac{dS}{dt} = 4t - 1$$

Note that the rate of change of distance with respect to time in expressions like 60 mph or 10 ft/sec are expressions for velocity. Since the derivative of distance is the rate of change of distance with respect to time, the equation $dS/dt = 4t - 1$ is the velocity at time $t \geq 0$. If we let v represent velocity, we have $v = 4t - 1$. Find the velocity of the object after 3 sec: $v = 4(3) - 1 = 11$ ft/sec.

Similarly, the rate of change of velocity with respect to time is acceleration. That is, acceleration a is given by $a = dv/dt$. Find the acceleration of the object in this example:

$$a = \frac{dv}{dt} = 4 \text{ ft/sec}^2$$

▶ **Exercise 3.3**

1 The distance d in feet is related to time t in seconds by $d = 16t^2$. Find the function that represents velocity. What is the velocity at $t = 2$ sec? Find the acceleration at $t = 3$ sec.

2 Let the profit P be related to the number n of units produced by $P = 200n - 2n^2$.
  a Find dP/dn (marginal profit).
  b Find the value of n that makes $dP/dn = 0$.
  c Find the values of n where dP/dn is positive.
  d Find the values of n where dP/dn is negative.
  e Graph the profit function.

3 If the length of a rectangle is 4 in. more than twice its width, then the area A of the rectangle is $A = w(2w + 4)$, where w is the width. Find the rate of change of area with respect to width.

4 Let cost (in thousands of dollars) be related to the number x of units produced by $C(x) = x^2 + 8x + 7$. Marginal cost is the rate of change of cost with respect to production, dC/dx.

a Find the marginal cost function.

b Graph the cost function.

c What is the fixed cost?

5 The area of a circle is given by the equation $A = \pi r^2$, where r is the radius of the circle. Find the function that yields the rate of change of the area with respect to the radius. Evaluate this function at $r = 5$.

6 As a science project, Mr. Peck's son rolled a boulder down a nearby mountain. The distance S in yards traveled was related to the time t in seconds by $S = 10t^{3/2}$.

a How far did the boulder roll in 16 sec? 25 sec?

b After 25 sec the boulder struck Mr. Peck's barn. What was the velocity of the boulder when it struck the barn?

c Was the velocity increasing at 25 sec?

7 A piece of cardboard 24 in. on each side is to be made into an open-top box. This is to be done by cutting a square of length x from each corner and folding up the flaps to form the sides. The volume of this box is given by $V = 576x - 96x^2 + 4x^3$.

a Find the rate of change of volume with respect to x.

b For what values of x is $dV/dx = 0$?

c What is the volume of the box for the value of x found in (b)?

8 Let profit P (in thousands of dollars) be related to the number n of units produced by $P = 4n - n^2$.

a What is the marginal profit function?

b What values of n make the marginal profit function (1) positive (2) zero (3) negative?

c Graph the profit function.

9 If an object is projected vertically upward with an initial velocity of 128 ft/sec, then its altitude S is given by $S = 128t - 16t^2$, where t is the time in seconds.

a Find the expression for the velocity of the object.

b After how many seconds will the velocity equal zero?

c How far will the object have traveled when the velocity reaches zero?

d Find the equation for the acceleration of the object.

10 Write the equation of the tangent line in each of the following.

a $y = x^2 + 2x + 3$ at (1, 6)

b $y = 10 - x^2$ at the point where $x = 2$

c $y = x^2 + 3x + 7$ at the point where $y = 5$

(There are two solutions to this problem.)

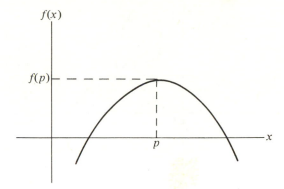

FIGURE 3.5

3.4 Locating Maximum and Minimum Points

In working with functions it is often useful to locate the largest functional value or the smallest functional value. In this section the derivative will be used to find the maximum and minimum values of the function. In order to concentrate on this application of the derivative, the functions will be restricted to those whose derivatives exist for every value of x in the domain of the function. The only restrictions on the domain of these functions will be mathematical restrictions.

DEFINITION

An *absolute maximum* occurs at the point $(p, f(p))$ if for $x = p$, $f(p)$ is greater than or equal to $f(x)$ for all other values of x in the domain of the function.

In Figure 3.5 $(p, f(p))$ is the absolute maximum point and $f(p)$ is the absolute maximum value.

DEFINITION

A *local maximum* occurs at the point $(p, f(p))$ if for $x = p$, $f(p)$ is greater than or equal to $f(x)$ for all values of x near p and in the domain of the function.

In Figure 3.6 $(p, f(p))$ is a local maximum point and $f(p)$ is a local maximum value. The value of $f(p)$ is not the absolute maximum value, because $f(b) > f(p)$.

DEFINITION

An *absolute minimum* occurs at the point $(p, f(p))$ if for $x = p$, $f(p)$ is less than or equal to $f(x)$ for all other values of x in the domain of the function.

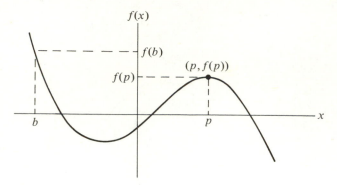

FIGURE 3.6

In Figure 3.7 $(p, f(p))$ is the absolute minimum point and $f(p)$ is the absolute minimum value.

DEFINITION

A *local minimum* occurs at the point $(p, f(p))$ if for $x = p$, $f(p)$ is less than or equal to $f(x)$ for all values of x near p and in the domain of the function.

In Figure 3.8 $(p, f(p))$ is a local minimum point and $f(p)$ is a local minimum value. The value of $f(p)$ is not the absolute minimum value, because $f(b) < f(p)$.

A function does not always have a local maximum or a local minimum. On the other hand, a function may have more than one local maximum or minimum. Finding these local maximum and minimum values, if they exist, is one important use of the derivative.

FIGURE 3.7

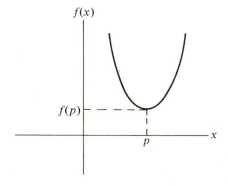

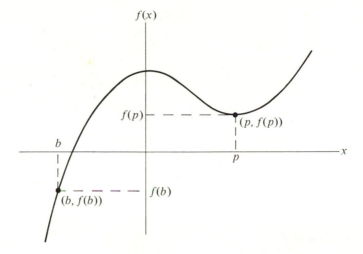

FIGURE 3.8

If a point on the graph of a function is a maximum point or a minimum point and the derivative exists at that point, then the tangent line is a horizontal line. (It does not matter whether the point is an absolute or a local maximum or minimum point.) Since the slope of every horizontal line is zero, the slope of the tangent line at a maximum or a minimum point is zero. Symbolically, this can be written as follows: If $(p, f(p))$ is either a maximum point or a minimum point, then $f'(p) = 0$. (See Figure 3.9.)

Thus the equation $f'(p) = 0$ does not specify whether $(p, f(p))$ is a maximum point or a minimum point. In fact, $f'(p) = 0$ may occur at a point where

FIGURE 3.9

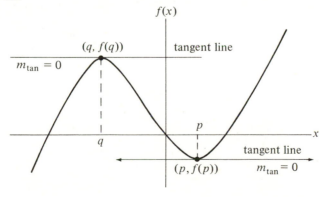

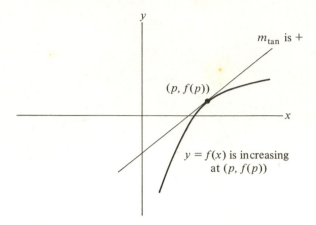

FIGURE 3.10

$(p, f(p))$ is neither a maximum point nor a minimum point. Therefore we must investigate more closely the behavior of the function in order to reach a definite conclusion about the possible maximum and minimum points.

If the slope of a tangent line to a curve is positive at a point on the curve, the function is said to be increasing at that point. (See Figure 3.10.)

If the slope of a tangent line to a curve is negative at a point on the curve, the function is said to be decreasing at that point. (See Figure 3.11.)

If the function is continuous on an interval around p, the point where the function changes from increasing (slopes of tangent lines positive) to decreasing (slopes of tangent lines negative) is a maximum point on the graph of the function. (See Figure 3.12.)

FIGURE 3.11

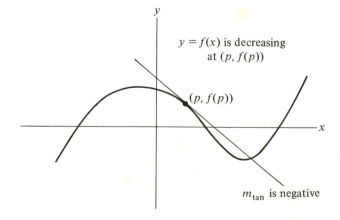

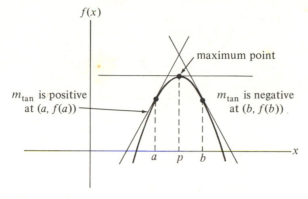

FIGURE 3.12

If the function is continuous on an interval around p, the point where the function changes from decreasing (slopes of tangent lines negative) to increasing (slopes of tangent lines positive) is a minimum point on the graph of the function. (See Figure 3.13.)

In order to use the derivative in finding maximum or minimum values of a nonconstant function, the value $x = p$ must be found such that $f'(p) = 0$. This can be done by setting $f'(x) = 0$ and solving the resulting equation for x. If there is no real-valued solution to the equation, there is no maximum or minimum value for the functions under consideration in this section. If there is a real-valued solution at $x = p$ (there may be more than one real-valued solution), then $f'(p) = 0$. If $f'(p) = 0$, then exactly one of the following three conditions must exist at $(p, f(p))$.

FIGURE 3.13

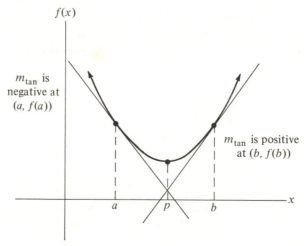

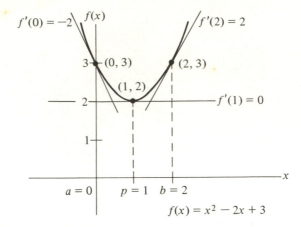

$f'(0) = -2$ $f(x)$ $f'(2) = 2$

3 (0, 3)   (2, 3)

(1, 2)

2 $f'(1) = 0$

1

x

$a = 0$ $p = 1$ $b = 2$

$f(x) = x^2 - 2x + 3$

FIGURE 3.14

1 $f(p)$ is a maximum value (either local or absolute).

2 $f(p)$ is a minimum value (either local or absolute).

3 $f(p)$ is neither a maximum nor a minimum value.

To determine which of the three possibilities occurs, select a value $x = a$ that is a little less than $x = p$. Then select a value $x = b$ that is a little more than $x = p$. Next, compute $f'(a)$ and $f'(b)$. If $f'(a)$ is positive and $f'(b)$ is negative, $(p, f(p))$ is a maximum point. If $f'(a)$ is negative and $f'(b)$ is positive, $(p, f(p))$ is a minimum point. If $f'(a)$ and $f'(b)$ are both positive, or both negative, $(p, f(p))$ is neither a maximum nor a minimum point.

There is a great deal of freedom in the selection of a and b. However, there are two restrictions that must be kept in mind. First, there must be exactly one solution to the equation $f'(x) = 0$ between the values $x = a$ and $x = b$. Second, the function must be continuous for all values of x, such that $a \leq x \leq b$.

Example 13 Let $f(x) = x^2 - 2x + 3$. Then $f'(x) = 2x - 2$. To find a value p such that $f'(p) = 0$, solve the equation $2x - 2 = 0$. Then $2x = 2$, and $x = 1$. If $a = 0$, then $f'(a) = f'(0) = 2(0) - 2 = -2$. If $b = 2$, then $f'(b) = f'(2) = 2(2) - 2 = 2$. Since $f'(0)$ is negative and $f'(2)$ is positive, $(1, f(1))$ is a minimum point on the graph of the function. Also, $f(1) = 2$, and 2 is the minimum value of the function. (See Figure 3.14.)

The fact that $(1, f(1))$ is a minimum point could have been established in the following manner: Compute $f(1)$, $f(0)$, and $f(2)$. These functional values would be 2, 3, and 3, respectively. Since $f(1)$ is less than $f(0)$ and $f(2)$, then $f(1) = 2$ is a minimum value.

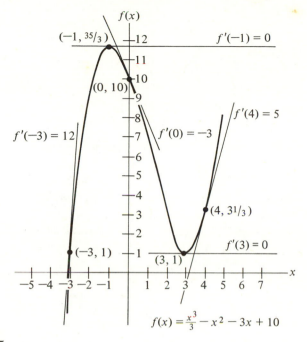

FIGURE 3.15

Example 14 Let $f(x) = x^3/3 - x^2 - 3x + 10$. Find all the maximum and minimum points for this function.

Now, $f'(x) = x^2 - 2x - 3 = 0$; so $(x - 3)(x + 1) = 0$. Then $x - 3 = 0$ and $x = 3$. Also, $x + 1 = 0$ and $x = -1$. Note that the solution to the derived equation has two values. These must be handled individually.

First, consider the case where $x = p = 3$. If $a = 0$, then $f'(a) = f'(0) = (0)^2 - 2(0) - 3 = -3$. If $b = 4$, then $f'(b) = f'(4) = (4)^2 - 2(4) - 3 = 5$. Since $f'(0)$ is negative and $f'(4)$ is positive, $(3, f(3))$ is a minimum point and $f(3) = 1$ is a minimum value of the function.

Second, consider the case where $x = p = -1$. If $a = -3$, then $f'(a) = f'(-3) = (-3)^2 - 2(-3) - 3 = 12$. If $b = 0$, then $f'(b) = f'(0) = (0)^2 - 2(0) - 3 = -3$. Since $f'(-3)$ is positive and $f'(0)$ is negative, $(-1, f(-1))$ is a maximum point and $f(-1) = 35/3$ is a maximum value of the function. (See Figure 3.15.)

A different method of determining the maximum and minimum points uses the functional evaluations. Compute $f(3), f(0)$, and $f(4)$. These functional values are 1, 10, and 10/3, respectively. Since $f(3)$ is less than $f(0)$ and $f(4)$, then $f(3) = 1$ is a minimum value. Then, compute $f(-1), f(-3)$, and $f(0)$. These functional values are 35/3, 1, and 10, respectively. Since $f(-1)$ is greater than $f(-3)$ and $f(0)$, then $f(-1) = 35/3$ is a maximum value.

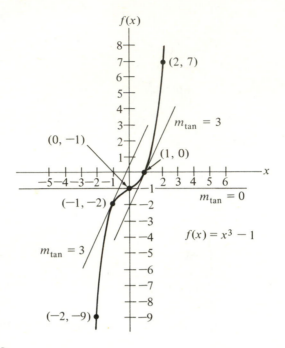

FIGURE 3.16

Example 15 Let $f(x) = x^3 - 1$. Find all the maximum and minimum points on the graph of this function.

Now, $f'(x) = 3x^2$. Solving $3x^2 = 0$, we have $x = 0$. If $a = -1$, then $f'(a) = f'(-1) = 3(-1)^2 = 3$. If $b = 1$, then $f'(b) = f'(1) = 3(1)^2 = 3$. Since $f'(-1)$ and $f'(1)$ are both positive, $(0, f(0))$ is neither a maximum nor a minimum point. (See Figure 3.16.)

We can also verify that $(0, f(0))$ is neither a maximum nor a minimum by computing $f(0)$, $f(-1)$, and $f(1)$. These functional values are -1, -2, and 0, respectively. Since $f(0)$ is between $f(-1)$ and $f(-2)$, then $f(0)$ is neither a maximum nor a minimum value of the function.

In Examples 13, 14, and 15, two methods were used to identify $(p, f(p))$ as a maximum point, a minimum point, or neither. (Remember that p was a solution of the derived equation $f'(x) = 0$.) The first method used in each case is called the *first derivative test*. The second method employed in each example is called the *functional values test*. The reader should compare the two tests.

Let $x = p$ be a solution for $f'(x) = 0$. Select $x = a$ such that a is a little less than p. Select $x = b$ such that b is a little more than p. (Recall that there are two limitations on this choice.) Table 3.3 summarizes the two methods for finding maximum and minimum values.

TABLE 3.3

	First derivative test	Functional values test
Compute	$f'(a)$ and $f'(b)$	$f(p)$, $f(a)$, and $f(b)$
Minimum	$f'(a) < 0$ and $f'(b) > 0$	$f(p) < f(b)$ and $f(p) < f(a)$
Maximum	$f'(a) > 0$ and $f'(b) < 0$	$f(p) > f(b)$ and $f(p) > f(a)$
Neither	$f'(a) > 0$ and $f'(b) > 0$ or $f'(a) < 0$ and $f'(b) < 0$	$f(p)$ is between $f(a)$ and $f(b)$

▶ **Exercise 3.4**

For each of the functions in Problems 1 through 15 find all of the maximum points and all of the minimum points. Note that the function may not have any maximum or minimum points. Use either test.

1 $f(x) = x^2 - 6x + 11$

2 $f(x) = \dfrac{x^3}{3} - 5x^2 + 21x + 10$

3 $f(x) = 10x - x^2$

4 $f(x) = 10{,}000 - 1500x + 90x^2 - x^3$

5 $f(x) = x^2 + 4x + 4$

6 $y = 90 + 40x - x^2$

7 $y = \dfrac{x^3}{3} - \dfrac{25x^2}{2} + 100x + 25$

8 $y = 2x^3 - 5$

9 $g(x) = x^2 - 30x + 250$

10 $g(x) = 6x + 5$

11 $f(x) = 50 - 200x + 15x^2 - \dfrac{x^3}{3}$

12 $y = \dfrac{1}{x}$

13 $y = 50x - x^2$

14 $f(x) = 1 + \dfrac{1}{x}$

15 $f(x) = 2x - \sqrt{x}$

16 For each of the Problems 1 through 15 that has a minimum value, determine whether it is an absolute minimum or a local minimum.

17 For each of the Problems 1 through 15 that has a maximum value, determine whether it is an absolute maximum or a local maximum.

3.5 Applications (Maximum and Minimum)

The techniques for finding maximum and minimum values of functions can be applied to such situations as finding the number of units of production to maximize profit, determining least average cost or maximum velocity, and so on. The following examples show some uses of this kind.

Applied Example 16 Suppose a rocket sled containing a fixed amount of fuel is to be driven down a test track. It has been determined that the velocity at any distance down the track (up to 10 km) can be represented by the equation $v = 500s - 50s^2$, where v is velocity in kilometers per hour and s is the distance in kilometers.

1. At what distance will the sled reach its maximum velocity?

$$v' = 500 - 100s = 0$$
$$500 = 100s$$
$$s = 5 \text{ km}$$

At $s = 4$ the velocity is increasing and at $s = 6$ the velocity is decreasing; therefore $s = 5$ is the distance at which the sled reaches its maximum velocity.

2. What is the maximum velocity?

$$v = 500(5) - 50(5)^2 = 1250 \text{ km/hr}$$

3. How far will the sled travel before it stops?

$$\text{At } v = 0, \qquad 0 = 500s - 50s^2$$
$$0 = s(500 - 50s)$$

Thus $s = 0$ and $s = 10$. Therefore the sled will have traveled 10 km.

Applied Example 17 (Quadratic Cost Function) Let the total production cost be represented by

$$C(x) = x^2 - 10x + 49$$

where x is the number of units produced.

1. Find the minimum of the cost function.

$$C'(x) = 2x - 10 = 0$$
$$x = 5$$

When $x = 5$, $C(5) = 24$. Since the slope of the function is negative for $x < 5$ and positive for $x > 5$, the point $(5, 24)$ is a minimum. Note that the cost must increase as production is increased. Therefore the domain of the cost function must be restricted to values of $x \geq 5$ for the function to be meaningful. The domains of all production cost functions are restricted to levels of production for which the cost is not decreasing.

2. The average cost $q(x)$ is defined to be $q(x) = C(x)/x$. Find the minimum of the average cost function.

$$q(x) = \frac{x^2 - 10x + 49}{x}$$

$$= x - 10 + \frac{49}{x}$$

$$q'(x) = 1 - \frac{49}{x^2} = 0$$

$$x^2 - 49 = 0$$

$$x = 7$$

(The solution $x = -7$ is not meaningful.) When $x = 7$, $q(7) = 4$. Since the slope of the function is negative for values of $x < 7$ and positive for values of $x > 7$, the point $(7, 4)$ is a minimum of the average cost function.

3. Find the marginal cost function.

$$C'(x) = 2x - 10$$

4. Show that the marginal cost is equal to the average cost at the production level where the average cost is a minimum, i.e., at $x = 7$.

We have $q(7) = 7 - 10 + 49/7 = 4$ and $C'(7) = 2(7) - 10 = 4$. Therefore $q(7) = C'(7)$. (This result can be proved in general by setting $q'(x) = 0$ and solving for x, then substituting the value of x into the equation $C'(x) = q(x)$.)

5. Graph the cost function, the marginal cost function, and the average cost function. (See Figure 3.17.)

Applied Example 18 (Revenue Function)   The revenue $R(x)$ obtained from the sale of x units, in thousands, is $R(x) = xp$, where p is the price per unit. If $p = 35 + 2x - x^2$, then $R(x) = 35x + 2x^2 - x^3$.

1. Find the maximum revenue.

$$R'(x) = 35 + 4x - 3x^2 = 0$$

$$x = \frac{4 \pm \sqrt{436}}{6}$$

$$x \doteq 4.13$$

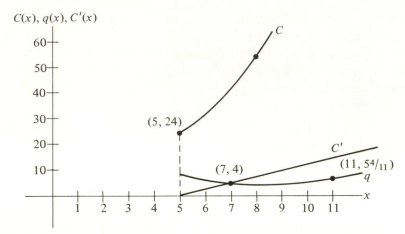

$C(x), q(x), C'(x)$

(5, 24)

(7, 4)

C'

$(11, {}^{54}/_{11})$

q

FIGURE 3.17

(The negative solution is not meaningful.) When $x \doteq 4.13$, $R(x) \doteq 105.9$. Checking the slopes on both sides of (4.13, 105.9) indicates that the point is a maximum.

2. The average revenue is defined to be $r(x) = R(x)/x$. Find the maximum average revenue.

$$r(x) = 35 + 2x - x^2$$
$$r'(x) = 2 - 2x = 0$$
$$x = 1$$

When $x = 1$, $r(1) = 36$. The slope is positive at $x = 0$, and the slope is negative at $x = 2$; therefore the point (1, 36) is a maximum.

3. Show that the marginal revenue is equal to the average revenue at the maximum of the average revenue function.

$$R'(x) = 35 + 4x - 3x^2$$
$$R'(1) = 35 + 4(1) - 3(1)^2 = 36$$
$$r(1) = 36, \text{ so}$$
$$r(1) = R'(1)$$

4. Graph the revenue function, the marginal revenue function, and the average revenue function. (See Figure 3.18.) Does the marginal revenue always have an x intercept at the value of x that corresponds to the maximum of the revenue function? Does the positive x intercept of the average revenue always correspond to the positive x intercept of the revenue?

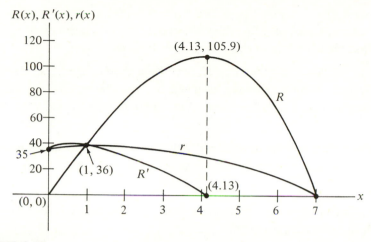

FIGURE 3.18

▶ **Exercise 3.5**

1 Let revenue R be related to the number of units produced, x, by $R(x) =$ $8x - x^2$. Find the maximum revenue.

2 A piece of cardboard is a square whose sides are 16 units each. A square of length x is cut from the corners of the cardboard and the flaps folded up to make an open-top box. Find the dimensions of the square cut from each corner that will maximize the volume of the box if the volume is $V =$ $x(16 - 2x)^2$. What is the maximum volume? (See Figure 3.19.)

FIGURE 3.19

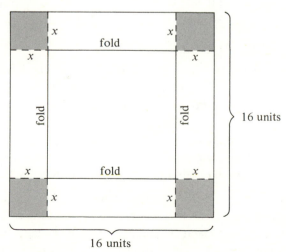

3 The cost $C(x) = 8x^2 - 12x + 128$, where x is the number of units produced.
  a Find the values of x for which the cost function is meaningful.
  b Find the minimum of the average cost function.

4 An airline has determined that its patronage in passenger miles depends on the rate r in cents per mile as given by $P = 600 - 100r$. Therefore the revenue $R = rP = r(600 - 100r)$. Find the rate r that yields the maximum revenue.

5 A plant manager has determined that the number of items n produced in an hour is related to the amount of coffee-break time t in minutes per hour by

$$n = 114 + 12t - t^2 \quad \text{for } 0 \le t \le 15$$

For what value of t is the maximum number of items produced? How many units are produced if $t = 0$?

6 A skyrocket is shot upward with an initial velocity of 80 ft/sec. The distance above ground is given by $s = 80t - 16t^2$, where t is in seconds and s is in feet.
  a What is the height of the skyrocket at $t = 1$?
  b What is the velocity of the skyrocket at $t = 1$?
  c When will the skyrocket reach its maximum height?
  d How high will the rocket go?

In Problems 7 through 9 find the average cost function and the marginal cost function. Show that the marginal cost is equal to the average cost at the minimum of the average cost function. Graph the cost function, the marginal cost function, and the average cost function.

7 $C(x) = x^2 + 10x + 4$ if $x \ge 0$

8 $C(x) = x^2 + 3x + 9$ if $x \ge 0$

9 $C(x) = x^2 - 8x + 25$ if $x \ge 4$

10 The revenue that is obtained from the sale of air conditioners is $R = px$, where x is the number of air conditioners sold and p is the price per air conditioner. The demand for air conditioners has determined that $p = 450 - 1.5x$.
  a Write R as a function of x.
  b Find the number of air conditioners sold that yields maximum revenue.
  c Find the price per air conditioner that will produce maximum revenue.

In Problems 11 through 13 (revenue functions) find the average revenue function and the marginal revenue function. Show that the marginal revenue is equal to the average revenue at the maximum of the average revenue function. Graph the revenue function, the marginal revenue function, and the average revenue function.

11 $R(x) = 20x + x^2 - x^3$

12 $R(x) = 30x + 4x^2 - x^3$

13 $R(x) = 10x + 2x^2 - 2x^3$

14 Let the number of items sold $S(d)$ be related to the number of dollars d (in hundreds) spent advertising by $S(d) = 500 + 60d - d^2$, where $0 \leq d \leq 40$. Find the amount spent for advertising that yields maximum sales. What is the maximum number of items that the company can expect to sell?

15 The profit P earned by the sale of x units produced and sold is given by $P = 50x - .01x^2 - 500$. What value of x yields a maximum profit? What is the maximum profit?

16 Let the cost $C = 60Q + 10,000$, where Q is the number of units produced weekly and 10,000 is in cents. The price is given by $p = 100 - .01Q$. (The price is in cents.) Find the number of units Q that will produce a maximum weekly profit. Find the maximum profit. Find the price at the value of Q that yields maximum profit.

17 Suppose the government imposes an eight-cent-per-unit tax on the company in Problem 16. Use the information in Problem 16 to determine the effect of the eight-cent tax. Find the number of units Q that will produce a maximum weekly profit. Find the maximum profit. Find the price when profit is maximized.

---

self-test • chapter three

1 Find the derivative of $f(x) = x^2 - 4$ by using the formula in the definition of the derivative.

In Problems 2 through 5 use the rules in order to find the derivative.

2 $f(x) = 40 + 3x - x^2$

3 $f(x) = 30/x^2$

4 $y = 4\sqrt{x} + 6$

5 $f(x) = 4x^3 - 3x^2 + 12$

6 Find the point on the graph of $f(x) = x^2 - 4x + 10$ where $f'(x) = 0$.

7 A cost function is given by the equation $C(x) = 3x^2 + 4x + 700$. Find the fixed cost, the marginal cost function, and the marginal cost at $x = 4$.

8 An object is projected vertically upward with an initial velocity of 256 ft/sec. The altitude s at any time t is given by the equation $s = -16t^2 + 256t$.
a Find the velocity after 1 sec.
b Find the distance the object has traveled after 2 sec.
c Find the maximum height the object will reach.

9 Let revenue be given by the equation $R(x) = 600x - \frac{1}{4}x^2$, where x is the number of units produced. Find the number of units produced that will yield the maximum revenue. Find the maximum revenue.

chapter four

the derivative with applications 2

4.1 The General Power Formula

In Chapter 3 some basic rules for finding derivatives were stated. Now we turn our attention to slightly more complicated functions and the rules for finding the derivatives of these functions. Some of the types of functions under consideration are the following:

$$f(x) = \frac{x - 2}{(x^2 + 3)^{1/2}}, \quad \text{a quotient of two quantities}$$

$$y = (x^2 + 1)^6, \quad \text{a quantity with a constant exponent}$$

$$g(x) = \sqrt{x + 1}(x^2 - 2x - 3), \quad \text{a product of two quantities}$$

$$f(x) = e^{3x-5}, \quad \text{an exponential } e \text{ function}$$

Rule 2 (Chapter 3) stated that if $y = x^n$, then $y' = nx^{n-1}$. In this section we would like to generalize in order to find the derivative when $y = [f(x)]^n$, where $f(x)$ is a function more complicated than $f(x) = x$. For example, $y = (2x^3 - 3)^5$, where $f(x) = 2x^3 - 3$. Rule 2 does *not* apply in this case. Rule 5 is needed.

RULE 5

If $y = [f(x)]^n$, where $f(x)$ is a differentiable expression, then $y' = n[f(x)]^{n-1} \cdot f'(x)$.

The derivative of an expression to an exponent is the exponent multiplied by the expression to the exponent minus 1, all multiplied by the derivative of the expression.

Proof (for n, a Counting Number) By definition,

$$y' = \lim_{x \to p} \frac{[f(x)]^n - [f(p)]^n}{x - p}$$

Factoring,

$$y' = \lim_{x \to p} \frac{[f(x) - f(p)][f(x)^{n-1} + f(x)^{n-2} \cdot f(p) + f(x)^{n-3} \cdot f(p)^2 + \cdots + f(p)^{n-1}]}{x - p}$$

$$= \lim_{x \to p} \frac{[f(x) - f(p)]}{x - p} \cdot \lim_{x \to p} [f(x)^{n-1} + f(x)^{n-2} \cdot f(p) + \cdots + f(p)^{n-1}]$$

$$= [f'(p)]\underbrace{[f(p)^{n-1} + f(p)^{n-1} + \cdots + f(p)^{n-1}]}_{n \text{ terms}}$$

$$= [f'(p)][nf(p)^{n-1}]$$

Replacing p by x

$$y' = n[f(x)]^{n-1} \cdot f'(x)$$

Example 1

1. If $y = (2x^3 - 3)^5$, then $y' = 5(2x^3 - 3)^4(6x^2)$.
2. If $y = \sqrt{2x + 1} = (2x + 1)^{1/2}$, then $y' = \frac{1}{2}(2x + 1)^{-1/2}(2) = 1/(2x + 1)^{1/2}$.
3. If $y = (x^2 - 5x)^{-2}$, then $y' = -2(x^2 - 5x)^{-3}(2x - 5) = (10 - 4x)/(x^2 - 5x)^3$.
4. If $C(n) = (2n - 7)^{-1/4}$, then $C'(n) = -\frac{1}{4}(2n - 7)^{-5/4}(2) = -1/(2(2n - 7)^{5/4})$.

▶ **Exercise 4.1**

Differentiate each of the following.

1 $y = (2x^2 + 3)^5 = 5(2x^2 + 3)^4(4x)$

2 $y = (5x^3 - 6x^2 + 7)^3 = 3(5x^3 - 6x^2 + 7)$

3 $y = (7 - 3x^2)^3 =$

4 $y = \sqrt[3]{x^3 - 6x}$

5 $y = (6x^2 + 7x - 8)^{-4}$

6 $y = (2x^2 - 7x + 1)^{4/3}$

7 $C = (2n - n^2)^5$

8 $R = \sqrt{2x + 3}$

9 $W = (q - 2q^4)^{-2}$

10 $g(x) = \dfrac{1}{(3x - 5)^2}$

11 $f(x) = (3x - 4)^{-1/2}$

12 $h(x) = (x^2 + 2x + 1)^{100}$

4.2   The Product Formula

RULE 6

If $f(x) = g(x) \cdot h(x)$, where $h(x)$ and $g(x)$ are differentiable functions, then $f'(x) = g(x) \cdot h'(x) + h(x) \cdot g'(x)$.

The derivative of a product of two differentiable expressions is the first expression multiplied by the derivative of the second expression plus the second expression multiplied by the derivative of the first expression.

Proof   By definition,

$$f'(p) = \lim_{x \to p} \frac{g(x) \cdot h(x) - g(p) \cdot h(p)}{x - p}$$

By using the mathematical technique of adding zero in the unusual form of $g(p) \cdot h(x) - g(p) \cdot h(x) = 0$, we have

$$f'(p) = \lim_{x \to p} \frac{g(x) \cdot h(x) - g(p) \cdot h(p) + g(p) \cdot h(x) - g(p) \cdot h(x)}{x - p}$$

Rearranging the terms,

$$f'(p) = \lim_{x \to p} \frac{g(x) \cdot h(x) - g(p) \cdot h(x) + g(p) \cdot h(x) - g(p) \cdot h(p)}{x - p}$$

Simplifying the fraction,

$$f'(p) = \lim_{x \to p} \left\{ \frac{[h(x)][g(x) - g(p)]}{x - p} + \frac{[g(p)][h(x) - h(p)]}{x - p} \right\}$$

Applying the properties of limits,

$$f'(p) = \left[\lim_{x \to p} h(x) \right] \left[\lim_{x \to p} \frac{g(x) - g(p)}{x - p} \right] + \left[\lim_{x \to p} g(p) \right] \left[\lim_{x \to p} \frac{h(x) - h(p)}{x - p} \right]$$

Therefore

$$f'(p) = h(p) \cdot g'(p) + g(p) \cdot h'(p)$$

Generalizing,

$$f'(x) = h(x) \cdot g'(x) + g(x) \cdot h'(x)$$

Example 2   Find the derivative of $f(x) = (x^2 + 2)(x + 3)$.

Then $f'(x) = (x^2 + 2)(1) + (2x)(x + 3)$. Simplifying, $f'(x) = 3x^2 + 6x + 2$. This derivative could have been found by computing the product first and then using the rules from Chapter 3. That is, $f(x) = (x^2 + 2)(x + 3) = x^3 + 3x^2 + 2x + 6$. Then $f'(x) = 3x^2 + 6x + 2$.

The product formula (Rule 6) is not needed if the product is easily obtained, but there are cases where it is necessary.

Example 3   Find the derivative of $y = \sqrt{x + 1}(x^2 + 3)$. Rewrite:

$$y = (x + 1)^{1/2}(x^2 + 3)$$

Then

$$y' = (x + 1)^{1/2}(2x) + (x^2 + 3)(\tfrac{1}{2})(x + 1)^{-1/2}(1)$$

Simplifying,

$$y' = (x + 1)^{1/2}(2x) + \frac{\frac{1}{2}(x^2 + 3)}{(x + 1)^{1/2}}$$

$$= \frac{2(x + 1)(2x) + (x^2 + 3)}{2(x + 1)^{1/2}}$$

$$= \frac{4x^2 + 4x + x^2 + 3}{2(x + 1)^{1/2}} = \frac{5x^2 + 4x + 3}{2\sqrt{x + 1}}$$

The amount of simplification required is determined by the intended use of the derivative. It may be better to limit the amount of simplification, as it greatly increases the chance of introducing computational errors.

Example 4 If $w = (q - 1)(2q - 6)$, then $w' = (q - 1)(2) + (2q - 6)(1)$. If $y = (x^2 + 6x + 3)(x^3 - 8)$, then $dy/dx = (x^2 + 6x + 3)(3x^2) + (x^3 - 8)(2x + 6)$.

▶ **Exercise 4.2**

Find the derivative of each of the following functions.

1 $y = (x^2 - 3)(2x + 5)$ 2 $y = \sqrt{x}(x^2 + 3)$

3 $y = x^6(x + 3)$ 4 $f(x) = (3x + 2)(5x - 4)$

5 $g(x) = \sqrt{x^2 - 1}(2x)$ 6 $y = 3x^{1/4}(x^2 + 2x + 1)$

7 $f(x) = (4x + 9)(3x^{-2} + 4x^{-1} + 6)$ 8 $y = (x + 5)^{1/2}(2x - 3)^{1/4}$

9 $y = (3x^4 - 2)^{1/2}(x^5 + 7x + 9)$ 10 $y = \sqrt{3x^4 - 1}(\sqrt[3]{2x^3 + x})$

11–12 Work Problems 1 and 4 by first simplifying and then using the rules in Chapter 3.

4.3   The Quotient Rule

RULE 7

If $f(x) = g(x)/h(x)$, where $g(x)$ and $h(x)$ are differentiable functions and $h(x) \neq 0$, then

$$f'(x) = \frac{h(x) \cdot g'(x) - g(x) \cdot h'(x)}{[h(x)]^2}$$

The derivative of the quotient of two functions is the denominator multiplied by the derivative of the numerator minus the numerator multiplied by the derivative of the denominator, all divided by the denominator squared.

The proof of this rule is left as an exercise.

Example 5   Find the derivative of $f(x) = (x^2 + 2)/(2x - 3)$. By Rule 7,

$$f'(x) = \frac{(2x - 3)(2x) - (x^2 + 2)(2)}{(2x - 3)^2}$$

Simplifying,

$$f'(x) = \frac{4x^2 - 6x - 2x^2 - 4}{4x^2 - 12x + 9} = \frac{2x^2 - 6x - 4}{4x^2 - 12x + 9}$$

Example 6   Find the derivative of $y = (2x + 1)/(x^2 - 3)$. Then

$$dy/dx = \frac{(x^2 - 3)(2) - (2x + 1)(2x)}{(x^2 - 3)^2}$$

If necessary, this derivative can be simplified, but extensive simplification is generally not necessary unless you want to let $f'(x) = 0$ and solve for possible maximum and/or minimum values.

In evaluating the derivative, replace x by p and then simplify.

Example 7   Let $y = (x^2 + 2x + 3)/(2x^2 - 6x + 1)$. Find the slope of the tangent line to the curve at $x = 0$. Write the equation of the tangent line to the curve at that point. Then

$$y' = \frac{(2x^2 - 6x + 1)(2x + 2) - (x^2 + 2x + 3)(4x - 6)}{(2x^2 - 6x + 1)^2}$$

At $x = 0$, $y' = [(1)(2) - (3)(-6)]/1 = 20$. When $x = 0$, $y = 3$. Thus the equation of the tangent line to the curve at the point $(0, 3)$ is $y - 3 = 20(x - 0)$. (The equation of the tangent line may be simplified to $y = 20x + 3$.)

▶ **Exercise 4.3**

Find the derivative of each of the functions in Problems 1 through 8.

1 $y = \dfrac{x^2}{3x - 2}$

2 $y = \dfrac{n^3}{n + 1}$

3 $g(x) = \dfrac{6x^2 + 2x}{x + 5}$

4 $P(n) = \dfrac{n^2 + 6n - 8}{n^2 - 5n + 4}$

5 $y = \dfrac{x^2 - 6x + 2}{2x + 5}$

6 $f(x) = \dfrac{5x^2 - 6x - 8}{x^3 - 7x^2 + 2}$

7 $g(x) = \dfrac{7x^2 + 7x}{x^3 + 5x}$

8 $y = \dfrac{(x + 1)^{1/2}}{x + 3}$

9–11 Find the slope of the tangent line to the curve in Problems 1, 2, and 3 at

$x = 1$. Write the equation of the tangent line at $x = 1$ for the first three problems.

12 Find the value of n where the tangent line is parallel to the n axis for the function $y = (n + 1)/n^2$.

13 Prove Rule 7.

4.4 The Derivative of $e^{g(x)}$

An important function encountered in calculus is $y = e^x$, where $e = \lim_{x \to \infty} (1 + 1/x)^x = 2.71828. \ldots$. The function $y = e^x$ is a very special function. It is equal to its derivative. This means that for each value of x the corresponding value of y and the slope of the tangent line to the curve at (x, y) are the same.

RULE 8 (Special Case)

If $y = e^x$, then $dy/dx = e^x$.

The function $y = e^x$ is equal to its derivative.

The proof of this rule requires the use of logarithms. However, the following argument is given to show the plausibility of this rule.

Argument By definition,

$$\frac{dy}{dx} = \lim_{x \to p} \frac{e^x - e^p}{x - p}$$

Factoring,

$$\frac{dy}{dx} = \lim_{x \to p} \frac{e^x(1 - e^{p-x})}{x - p}$$

By Theorem 4 of limits,

$$\frac{dy}{dx} = \lim_{x \to p} e^x \cdot \lim_{x \to p} \frac{1 - e^{p-x}}{x - p}$$

The

$$\lim_{x \to p} \frac{(1 - e^{p-x})}{x - p}$$

can be indicated by the tabular method. (See Table 4.1.) The concept of $x \to p$ is the same as $x - p \to 0$.

TABLE 4.1

Value of $x - p$	Value of $\dfrac{1 - e^{p-x}}{x - p}$
As $x \to p$ from the right	
.5	.787
.3	.864
.1	.952
.01	.995
.001	.9995
As $x \to p$ from the left	
$-.5$	1.3
$-.3$	1.166
$-.1$	1.052
$-.01$	1.00501
$-.001$	1.00005

Thus

$$\lim_{x \to p} \frac{1 - e^{p-x}}{x - p} = 1$$

Since $\lim_{x \to p} e^x = e^p$, we have $dy/dx = e^p \cdot 1 = e^p$. Replacing p by x, $dy/dx = e^x$.

Example 8   Let $y = e^x$. Find y'. By the special case of Rule 8, $y' = e^x$. If $x = 0$, then $y = e^0 = 1$ and $y' = e^0 = 1$. The value of y and the slope of the tangent line are the same.

When the function is $y = e^{g(x)}$, where $g(x)$ is a differentiable function of x (other than $g(x) = x$), the property of the function being equal to its derivative is lost.

RULE 8

If $y = e^{g(x)}$, then $y' = g'(x)e^{g(x)}$.

The derivative of e to an exponent is e to that exponent multiplied by the derivative of that exponent.

Example 9

$$\text{If } y = e^{2x}, \text{ then } y' = 2e^{2x}$$
$$\text{If } y = e^{x^3}, \text{ then } y' = 3x^2 e^{x^3}$$
$$\text{If } y = e^x, \text{ then } y' = (1)e^x = e^x$$

Hence Rule 8 also applies to the special case $g(x) = x$.

A function of the form $f(x) = e^{g(x)}$ occurs in applied problems of various disciplines. Some of these applications will be identified in the following examples and problems.

Applied Example 10 An altimeter measures altitude by measuring atmospheric pressure. This is possible because pressure decreases as height above sea level increases. If the temperature remains constant, the pressure p in inches of mercury is related to the height h in miles by $p = 30e^{-.23h}$.

1. Find the height above sea level when the atmospheric pressure is 7.5 in. of mercury.

We have $7.5 = 30e^{-.23h}$; then $.25 = e^{-.23h}$. Let $x = .23h$; then $.25 = e^{-x}$. Consulting the table in Appendix 3, we have $1.4 = x$. Therefore $1.4 = .23h$, and $h = 1.4/.23 = 6.09$ miles.

2. Find the expression for the rate of change of atmospheric pressure with respect to height.

$$\frac{dp}{dh} = (30)(-.23e^{-.23h})$$

$$= -6.90(e^{-.23h})$$

3. Find the rate of change of pressure with respect to height when $h = 1$ mile.

$$\frac{dp}{dh} = -6.90(e^{-.23h})$$

$$= -6.90(.8) = -5.52 \text{ in. mercury/mile}$$

4. Since $e^{-.23h}$ is always positive, $dp/dh = -6.90(e^{-.23h})$ is always negative. Therefore the function is always decreasing.

▶ **Exercise 4.4**

Find the derivative of each of the following functions.

1 $y = e^x + 3$ 2 $y = 2e^x$

3 $y = -\frac{2}{3}e^x$ 4 $f(x) = \sqrt{2}(e^{-x})$

5 $y = e^{3x}$ 6 $g(x) = e^{-3x}$

7 $s = e^{t^2}$ 8 $q(n) = 3e^{3n}$

9 $y = \sqrt{2}(e^{\sqrt{2}x})$ 10 $R(x) = e^{x^4}$

11 $y = e^{2x^2-6x}$ 12 $w = 2e^{2x-1}$

13 $y = xe^x$ 14 $y = \dfrac{x^2}{e^x}$

15 $f(x) = xe^{x^2-1}$ 16 $f(x) = (x^2 + 1)e^{2x}$

17 $g(x) = \dfrac{e^x}{x^2}$

18 $f(x) = (e^x)^3$

19 $y = x^2 + 1 + e^x$

20 $f(x) = \dfrac{1 - e^x}{x}$

21 For a certain temperature, the atmospheric pressure is related to the height in miles by $P = 30(e^{-.2h})$. Find the height when $P = 10$. Find the rate of change of P with respect to h when $h = 2$ miles.

22 With the supply of gasoline decreasing, the government decided to increase the tax on gasoline in order to decrease consumption from the present 10^6 gal/day. The consumption c of gasoline is related to the increase in tax per gallon x by the equation $c = 10^6(e^{-.01x})$.

  a How much gasoline will be consumed per day if the tax is increased by 10 cents per gallon?

  b What is the percentage decrease in consumption?

  c Find the expression for the rate of change of consumption with respect to x.

  d Is the consumption of gasoline decreasing at $x = 2$?

  e What is the rate of change of consumption at $x = 10$?

23 A peaceful spring-fed creek normally flows at 12 ft above sea level. When it rains in the creek's watershed, the level of the creek increases rapidly, approaching a flood level of 32 ft. The level of the creek L is related to the number of inches of rain r in its watershed as given by $L = 20(1 - e^{-.2r}) + 12$.

  a To what level does the creek rise after 2 in. of rainfall in its watershed?

  b At what rate is the level of the creek increasing after 2 in. of rainfall in its watershed?

  c Is the function in this problem increasing at $r = 3$?

24 The amount A of radiant energy absorbed by a material depends on the thickness of the material t, the intensity of the radiation source R, and a positive constant c called the coefficient of absorption, as given by $A = Re^{ct}$. If the coefficient of absorption is .4 for a particular substance and R is constant, find the rate of change of energy absorbed with respect to the thickness. What is the rate of change of energy absorbed at $t = 5$? Why is the function $A = Re^{.4t}$ always increasing?

25 A geophysicist sometimes finds it necessary to compute the velocity of a shock wave passing through the earth. However, shock waves travel at different velocities at different depths below the surface. One equation used to determine such velocities is $v = v_0 e^{kx}$, where x is the depth of the shock wave in miles, v_0 is the velocity of the shock wave at depth zero, v is the velocity of the shock wave at depth x, and k is a constant ranging from .5 to 1.5, depending on the soil in the region.

a If v_0 is .2 miles/sec and $k = .5$, find the velocity of a shock wave traveling 2 miles deep.

b Find the expression for rate of change of velocity with respect to depth.

c Find the rate of change of velocity at a depth of 2 miles.

26 The selling price S of a given item is related to the number of units x available for sale by $S = 5e^{-.001x}$. Find the selling price if 1000 units are available. Find the rate of change of selling price with respect to the number of units. Is the selling price decreasing?

▶ Summary Exercise 4.4

Find the derivative of each function. Be prepared to use Rules 1 through 8 and combinations of them.

1 $y = (2x^4 - 6x^2 + 8)^5$

2 $y = (2x^2 + 3x)(x - 4x^5)$

3 $y = e^{3x-2}$

4 $Q(n) = \dfrac{n^2 + n}{n + 5}$

5 $y = (x^3 + 2x)(5x - 1)^3$

6 $y = x^3 e^{x^2}$

7 $y = \dfrac{x}{e^x}$

8 $y = \dfrac{e^{-x}}{e^{x^2} + x}$

9 $y = \dfrac{\sqrt{2x + 5}}{\sqrt{x^2 + 1}}$

4.5 More Applications of the Derivative

In this section the eight rules of differentiation will be used to maximize and minimize different functions. These problems are similar to the maximum and minimum problems in Chapter 3, with two distinct differences. First, Rules 5 through 8 may be needed in order to find the derivative. Second, in some of the problems the student will have to build the equation. It is suggested that you study the following examples carefully before starting to translate words into correct mathematical equations.

Applied Example 11   A restaurant manager has 100 places in his restaurant. He knows from experience that each place yields an average income of $25/day. In planning to enlarge the restaurant, he estimates that for each additional place, his average income per place will decrease by 10 cents. How many places should be added in order to maximize the average daily income of the restaurant?

At present, $R = (100)(25) = \$2500$. If 20 places are added, then $R = (100 + 20)[25 - 20(.10)] = \$2760$. If 120 places are added, then $R = (100 + 120)[25 - 120(.10)] = \$2860$. More generally, let x be the number of places added. Then $R = (100 + x)[25 - x(.10)]$. $R' = (100 + x)(-.10) +$

$(25 - .10x)(1)$. Then $0 = -10 - .10x + 25 - .10x$, and $.20x = 15$, and $x = 75$. If $x = p = 75$, then $R(p) = R(75) = (100 + 75)[25 - .10(75)] = \$3062.50$.

To determine if \$3062.50 is actually the maximum expected value, one of the tests for maximum and minimum points can be used. If $x = a = 20$, then $R(20) = \$2760$. If $x = b = 120$, then $R(120) = \$2860$. Since \$3062.50 is greater than \$2760 and \$2860, then $R(75) = \$3062.50$ is a maximum value. Therefore the restaurant should add 75 places in order to maximize revenue.

Applied Example 12 A man wants to enclose a rectangular area with a fence. One side of the area will be an existing building. The fence opposite the building will cost \$2/ft, while the two side fences will cost \$1/ft. If he wants to enclose an area of 10,000 sq ft, what dimensions should he use to minimize cost?

Let x be the length of one side fence and y be the length of fence opposite the building. The area is $xy = 10,000$. (See Figure 4.1.)

In order to minimize cost, there must be a cost function. Thus $C = 2y + 2(1)x$. The cost of each side is the cost per foot multiplied by the number of feet. In order to minimize cost, the cost function must contain only one independent variable. Therefore y must be replaced by some equivalent expression containing only constants and the variable x. Since $xy = 10,000$, then $y = 10,000/x$. Then cost can be written as a function of a single variable x by replacing y with $10,000/x$.

$$C(x) = 2\left(\frac{10,000}{x}\right) + 2(1)x = 20,000x^{-1} + 2x$$

Differentiating,

$$C'(x) = -20,000x^{-2} + 2$$

Setting the derivative equal to zero and solving,

$$\frac{-20,000}{x^2} + 2 = 0$$

$$2x^2 = 20,000$$

$$x^2 = 10,000$$

$$x = 100 \quad \text{or} \quad x = -100$$

FIGURE 4.1

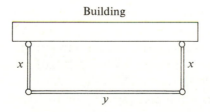

Building

x x

y

We can discard $x = -100$, since -100 is not in the domain of the function. (Why?)

Test:

$$\text{If } x = p = 100, \text{ then } C(p) = C(100) = 400.$$
$$\text{If } x = a = 10, \text{ then } C(a) = C(10) = 2020.$$
$$\text{If } x = b = 10{,}000, \text{ then } C(b) = 20{,}002.$$

In comparing $C(p)$, $C(a)$, and $C(b)$, we see that $C(p)$ is the smallest. Therefore a minimum cost of $400 is produced when $x = 100$. Since $y = 10{,}000/x$, $y = 100$. Hence the dimensions are 100 by 100 ft.

In solving a stated problem the following procedure might be helpful.

1 Label the unknowns. There may be more than one unknown in the beginning.

2 If possible, draw a figure.

3 Determine what is to be maximized or minimized (cost, area, volume, profit, and so on).

4 Write a function to represent the quantity determined in step 3 (cost function, area function, volume function, profit function, and so on). At this point the function may have more than one independent variable.

5 If the function in step 4 has more than one independent variable, there must be some given fact that will allow equivalent replacements for all independent variables except one.

6 Find the maximum or minimum value of the function in step 5.

7 Use the value found in step 6 to compute the corresponding values for the variables of step 1.

▶ **Exercise 4.5**

1 Suppose that the selling price for a certain item is determined by the number of units n for sale. The selling price S is given by $S = 10e^{-.005n}$.
 a Find the total revenue function.
 b Find the number of units n that maximizes the revenue.
 c Find the selling price S that corresponds to the value of n found in (b).

2 A pool is being drained for cleaning. The volume v in gallons remaining in the pool at any time t in hours is given by $v = 100(50 - t)^2$.
 a What is the rate of change of volume with respect to time?
 b What is the volume in the pool when it is full?
 c At what time will the rate of change of volume be zero?
 d How many hours will it take for the pool to drain?

3 A firm sells all units it produces at $10/unit. The firm's total cost of producing x units is $C(x) = 100 + 1.5x + .01x^2$.
a Write the equation of the profit function.
b From the result in (a), find the number of units that should be produced to maximize profit.
c Find the maximum profit that the firm can expect.

4 The price p for which an item can be sold is a function of the amount x available for sale, as given by $p = 2e^{-x/100}$. Assuming that all available units are sold, write the revenue function. How many units should be marketed in order to maximize the revenue?

5 A rectangular area of 600 sq yd is to be enclosed by a fence, then partitioned into two parts by a fence down the middle (parallel to the length). Find the minimum amount of fence needed to enclose the area. (*Hint:* The length of fence is $3L + 2W$.)

6 A rectangular area is enclosed by a fence and divided down the middle by another fence (parallel to the length). If 2400 ft of fence is used, find the maximum area that can be enclosed.

7 A box with a square top and bottom contains a volume of 216 cu in. What are the dimensions of the box if the surface area is a minimum? What is the minimum surface area of the box?

8 A rectangular area of $11,666\frac{2}{3}$ sq ft is to be enclosed by a fence and divided down the middle by another fence (parallel to the length). The fence down the middle costs $.50/ft and the rest of the fence costs $1.50/ft.
a Write the cost function for the fence.
b Find the dimensions that yield minimum cost.
c What is the minimum cost of the fence?

9 If 25 lemon trees are planted per acre in a certain orchard, the yield will be 525 lemons per tree. For each additional tree planted per acre the yield per tree will be reduced by 15 lemons. What is the optimum number of trees to plant per acre?

10 An open-top box with a square base and a volume of 64 cu ft is to be made from thin sheet metal. Find the dimensions for minimum cost if the bottom costs $2.00/sq ft and the sides cost $1.00/sq ft.

11 A restaurant manager knows from experience that each of his 175 places yields an average income of $25 per day. He estimates that for each place he adds to the restaurant his average daily income will be reduced by 4 cents per place. How many places should be added to achieve maximum income?

12 A box is to be made from a piece of cardboard 32 in. sq by cutting equal squares from the corners and turning up the sides. Find the dimensions of the largest box that can be so made.

13 A rectangular container with an open top is to be made from a piece of tin that is 15 in. long and 8 in. wide by cutting equal squares from each corner and turning up the sides. If the square is of length x, write the volume of the container as a function of x. Find the dimensions that maximize the volume of the container.

4.6 Curve Sketching

In graphing a function that is not a linear function, two important characteristics of the graph that must be determined are its continuity and its direction (increasing or decreasing).

In determining continuity, one must find the domain of the function. Mathematical restrictions are vital in finding the domain.

Example 13 Let $f(x) = 1/(x - 1)$. This function is undefined at $x = 1$, since $f(1) = 1/0$. Since the denominator is zero at $x = 1$, there can be no point on the graph of the function with an x value of 1. Since the function is undefined at $x = 1$, a dotted vertical line is drawn on the graph at $x = 1$. (See Figure 4.2.) This line is called a *vertical asymptote*. The vertical asymptote serves to divide the graph of the function into distinct parts, one to the left of the line and the other to the right of the line.

Many functions have a horizontal asymptote. In Figure 4.2 the graph approaches the x axis (whose equation is $y = 0$) as x approaches infinity or negative infinity. For a rational function the equation of the horizontal asymptote is the limit of the function as x approaches infinity. This is denoted by

FIGURE 4.2

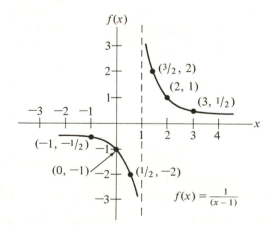

$\lim_{x\to\infty} f(x) = L$. If L is finite, then $y = L$ is the equation of the horizontal asymptote.

Example 14 Find the equation of the horizontal and vertical asymptotes, if there are any, for $f(x) = (2x + 3)/(3x - 4)$. Since $3x - 4 = 0$ is not allowed, $x = \frac{4}{3}$ is a vertical asymptote. Then $\lim_{x\to\infty} [(2x + 3)/(3x - 4)] = \frac{2}{3}$, and $y = \frac{2}{3}$ is the equation of the horizontal asymptote.

Next we must determine the values of x for which the function is increasing and the values of x for which the function is decreasing. The graph of a function changes direction in one of two ways. First, it changes from increasing to decreasing at a maximum point and from decreasing to increasing at a minimum point. Second, it may change direction at a point of discontinuity. (See Figures 4.3 and 4.4.)

To sketch the graph of a function, the following procedure may be used.

1 Find all points of discontinuity. If $f(p)$ is undefined, a vertical asymptote will usually occur at $x = p$.

2 Evaluate $\lim_{x\to\infty} f(x)$ and $\lim_{x\to-\infty} f(x)$. If either limit is finite and equal to L, then $y = L$ is the equation of a horizontal asymptote.

3 Find all of the maximum and minimum points of the function. Plot these points.

4 Determine the direction of the curve between each x value found in steps 1 and 3.

5 Sketch the curve. The sketch is more likely to be accurate if a few points are found. Usually, these points should have x values that are near the x values found in steps 1 and 3. Also, the x and y intercepts are good points to plot.

FIGURE 4.3

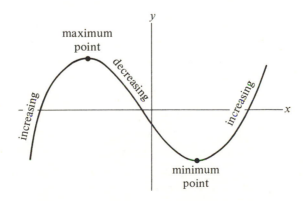

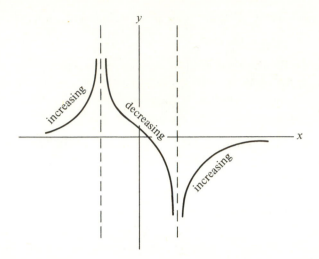

FIGURE 4.4

However, in some functions the x intercept is difficult, if not impossible, to find.

Example 15 Sketch the graph of the following function: $f(x) = x^3/3 - 3x^2 + 5x + 2$.

1. There are no values of x that make the function undefined. Therefore, there is no vertical asymptote.

2.

$$\lim_{x \to \infty} \left(\frac{x^3}{3} - 3x^2 + 5x + 2 \right) = \infty$$

and

$$\lim_{x \to -\infty} \left(\frac{x^3}{3} - 3x^2 + 5x + 2 \right) = -\infty$$

Therefore there is no horizontal asymptote.

3. Since $f(x) = x^3/3 - 3x^2 + 5x + 2$, then $f'(x) = x^2 - 6x + 5$. Set $x^2 - 6x + 5 = 0$. Factoring yields $(x - 5)(x - 1) = 0$, so $x = 5$ and $x = 1$. Test $x = 5$ and $x = 1$ for maximum or minimum values. At $x = 1$, $f(1) = 4\frac{1}{3}$. Let $a = 0$. Then $f(0) = 2$. Let $b = 2$. Then $f(2) = 2\frac{2}{3}$. Thus $(1, 4\frac{1}{3})$ is a maximum point of the graph of the function.

At $x = 5$, $f(5) = -6\frac{1}{3}$. Let $a = 2$. Then $f(2) = 2\frac{2}{3}$. Let $b = 6$. Then $f(6) = -4$. Thus $(5, -6\frac{1}{3})$ is a minimum point on the graph of the function.

4. The function is increasing for all values of x less than 1. The function is

decreasing for all values of x between 1 and 5. The function is increasing for all values of x greater than 5.

5. Since the functional values test was used to determine the maximum and minimum points, we have some points on the graph already computed, namely $(0, 2)$, $(6, -4)$, and $(2, 2\frac{2}{3})$. (Refer to step 3.) (See Figure 4.5.)

Example 16   Sketch the graph of $f(x) = 1/(x^2 - 1)$.

1. The function is discontinuous when $x^2 - 1 = 0$. Solving $x^2 - 1 = 0$ yields $x = 1$ and $x = -1$. Since $f(x) = 1/(x^2 - 1)$ is undefined at $x = 1$ and at $x = -1$, there are two vertical asymptotes. One of them is at $x = 1$ and the other is at $x = -1$.

2.

$$\lim_{x \to \infty} \frac{1}{x^2 - 1} = 0 \quad \text{and} \quad \lim_{x \to -\infty} \frac{1}{x^2 - 1} = 0$$

Thus $y = 0$ is the equation of the horizontal asymptote. (See Figure 4.6.)

3. Find any maximum and/or minimum points on the graph. Since $f(x) = 1/(x^2 - 1)$,

$$f'(x) = \frac{(x^2 - 1)0 - 1(2x)}{(x^2 - 1)^2} = \frac{-2x}{(x^2 - 1)^2}$$

Set $-2x/(x^2 - 1)^2 = 0$. To solve the equation, take the numerator of the fraction and set it equal to zero. The solution obtained must be checked in the denominator in order to make sure that the solution does not make the denominator equal to zero. If the solution does make the denominator equal to zero, then the solution is not valid and must be discarded.

FIGURE 4.5

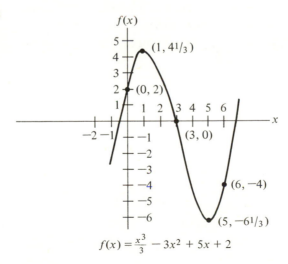

$f(x) = \frac{x^3}{3} - 3x^2 + 5x + 2$

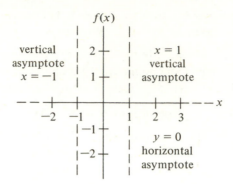

FIGURE 4.6

Set $-2x = 0$. Solving for x yields $x = 0$. Test $x = 0$ for maximum or minimum values. At $x = 0$, $f(0) = -1$. Let $a = -\frac{1}{2}$. Then $f(-\frac{1}{2}) = -\frac{4}{3}$. Let $b = \frac{1}{2}$. Then $f(\frac{1}{2}) = -\frac{4}{3}$. Therefore $(0, -1)$ is a maximum point on the graph of the function. Note that the value for a must be between $x = 0$ and $x = -1$ and the value for b must be between $x = 0$ and $x = 1$. The values of a and b must be restricted, because the function is discontinuous at $x = -1$ and at $x = 1$. Also, $(-\frac{1}{2}, -\frac{4}{3})$ and $(\frac{1}{2}, -\frac{4}{3})$ are points on the graph of the function. (See Figure 4.7.)

The two vertical asymptotes indicate that the graph of this function is divided into three distinct parts. The absence of maximum or minimum points in the other two parts indicates that the graph of the function does not change direction to the left of the vertical asymptote whose equation is $x = -1$; nor does it change direction to the right of the vertical asymptote whose equation is $x = 1$. To find the direction of the graph in each of these two parts, compute the functional values for two values of x in each interval and then compare the functional values. In one part, select $x = -3$ and $x = -2$. At $x = -3$, $f(-3) = \frac{1}{8}$. At $x = -2$, $f(-2) = \frac{1}{3}$. Since $\frac{1}{3} > \frac{1}{8}$, the graph of the function is

FIGURE 4.7

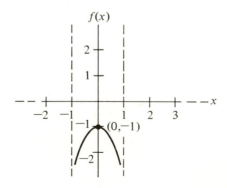

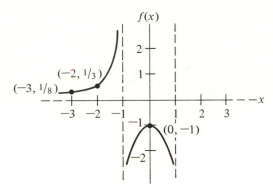

FIGURE 4.8

increasing in the part of the graph to the left of $x = -1$. Plot the two points that have just been determined. Remember that $y = 0$ is a horizontal asymptote and $x = -1$ is a vertical asymptote. (See Figure 4.8.)

In the remaining part of the graph, select $x = 2$ and $x = 3$. At $x = 2$, $f(2) = \frac{1}{3}$. At $x = 3$, $f(3) = \frac{1}{8}$. Since $\frac{1}{3} > \frac{1}{8}$, the graph is decreasing in the part of the graph to the right of $x = 1$. (See Figure 4.9.)

Example 17   Sketch the graph of $f(x) = e^{-x^2}$.

1. Since $e^{-x^2} = 1/e^{x^2}$, set $e^{x^2} = 0$. There is no solution to this equation since $e^{x^2} > 0$ for all values of x. Hence $f(x) = e^{-x^2}$ is continuous for all values of x and there are no vertical asymptotes.

FIGURE 4.9

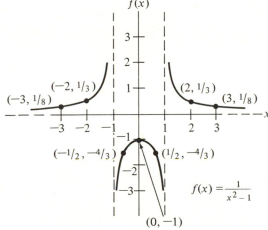

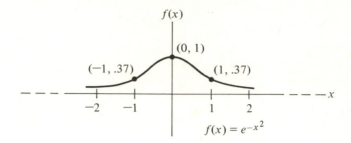

FIGURE 4.10

2. Evaluate

$$\lim_{x \to \infty} e^{-x^2} = \lim_{x \to \infty} \frac{1}{e^{x^2}}$$

As $x \to \infty$, $e^{x^2} \to \infty$. (Recall that $e > 1$.) Therefore as $e^{x^2} \to \infty$, $1/e^{x^2} \to 0$. Restating, $\lim_{x \to \infty} 1/e^{x^2} = 0$. When e to some variable power is being examined, it is also necessary to examine $\lim_{x \to -\infty} 1/e^{x^2}$. Since $e^{x^2} \to \infty$ as $x \to -\infty$, then $\lim_{x \to -\infty} 1/e^{x^2} = 0$. The graph is asymptotic to the x axis as $x \to \infty$ and as $x \to -\infty$. Thus $y = 0$ is the equation of the horizontal asymptote. (A function involving e to a variable exponent is generally not asymptotic at both ends of the graph.)

3. Since $f(x) = e^{-x^2}$, then $f'(x) = e^{-x^2}(-2x)$. Set $e^{-x^2} = 0$. This equation does not have a solution, since e to any exponent is always greater than zero. Set $-2x = 0$. Then $x = 0$.

Test:

If $x = p = 0$, then $f(p) = f(0) = 1$.

If $x = a = -1$, then $f(a) = f(-1) = .37$.

If $x = b = 1$, then $f(b) = f(1) = .37$.

Since 1 is greater than .37, then $(0, 1)$ is a maximum point on the graph of the function. (See Figure 4.10.)

▶ **Exercise 4.6**

Find the equation of the vertical asymptotes and/or the equation of the horizontal asymptotes for each of the functions in Problems 1 through 6. It is not certain that a particular function has either a vertical asymptote or a horizontal asymptote.

1 $f(x) = \dfrac{1}{x}$

2 $f(x) = \dfrac{3x^2 + 4}{2x^2 - 3x - 5}$

3 $y = \dfrac{4x}{x^2 + 1}$

4 $f(x) = \dfrac{x^2}{2x - 1}$

5 $f(x) = \dfrac{5x^2 - 3x + 20}{x^2 - 9}$

6 $y = \dfrac{x^3 + 1}{x + 1}$

Sketch the graph of each of the functions in Problems 7 through 18.

7 $f(x) = x^2 - 3x + 5$

8 $f(x) = -2x^2 + 8x + 3$

9 $f(x) = -x^2 + 6x$

10 $f(x) = x^3 - 3x$

11 $f(x) = \dfrac{x^3}{3} + \dfrac{x^2}{2} - 2x$

12 $f(x) = x^4 - 2x^2 + 3$

13 $f(x) = \dfrac{2x}{x - 1}$

14 $f(x) = \dfrac{6}{x^2 - 25}$

15 $f(x) = \dfrac{x^2}{x + 2}$

16 $f(x) = e^{x^2}$

17 $f(x) = 3e^{-x}$

18 $f(x) = e^{-3x^2}$

---

self-test • chapter four

In Problems 1 through 4, find the derivative of each function.

1 $y = \sqrt{2x^4 + 6x}$

2 $y = \dfrac{x^2 + 3}{3x - 2}$

3 $y = e^{-3x+1}$

4 $y = xe^{x^2}$

5 Let $f(x) = 3/(x - 2)$.
   a Find the equation of the vertical asymptote(s), if any.
   b Find the equation of the horizontal asymptote(s), if any.
   c Find all the local maximum and minimum points.
   d Graph the function.

6 Let $f(x) = x^3 - 5x^2 + 3x + 1$. Repeat steps **a**–**d** of Problem 5.

7 A dinner theater has 100 places. The manager knows that his average weekly income is \$75/place. He estimates that each additional place that he adds in a planned expansion will reduce the average weekly income by 50 cents/place. How many places should he add in order to have maximum income?

8 Let x be the number of items. The selling price for this item is given by $p = 50 - .02x$.
   a Find the revenue function.
   b Find the value of x that will produce maximum revenue.

chapter five

the indefinite integral

5.1 The Concept of Integration

In calculus, as in arithmetic, there are operations and their opposites. For example, when given factors are multiplied, we obtain a single (unique) answer called the *product*. In the opposite operation, called *factoring*, the product is given and the factors are desired.

Example 1

Multiplication	*Factoring*
$(3)(2) = 6$	$6 = (3)(2)$
$(x + 1)(x - 3) = x^2 - 2x - 3$	$x^2 - 2x - 3 = (x + 1)(x - 3)$
$x^2(x + 1) = x^3 + x^2$	$x^3 + x^2 = x^2(x + 1)$

However, factoring does not give a single (unique) result (unless prime factors are specified).

Example 2

Multiplication

$(9)(6) = 54$

Factoring

$$54 = (9)(6)$$
$$= (27)(2)$$
$$= (-3)(-18), \text{ etc.}$$

$(x)(x + 1)(x + 2) = x^3 + 3x^2 + 2x$

$$x^3 + 3x^2 + 2x = x(x^2 + 3x + 2)$$
$$= (x^2 + x)(x + 2), \text{ etc.}$$

In calculus a similar relation exists between the operations of integration and differentiation. When we differentiate, we obtain a single (unique) result called the *derivative*.

Example 3

Function	Derivative
$y = 6x^3 - 2x + 3$	$y' = 18x^2 - 2$
$C(n) = n^2 - 6n + 50$	$C'(n) = 2n - 6$
$g(x) = e^{3x}$	$g'(x) = 3e^{3x}$

In Example 4 all the functions differ by a constant, but the derivatives are the same because the derivative of the constant is zero.

Example 4

Function	Derivative
$y = 2x^2 + 6x$	$y' = 4x + 6$
$y = 2x^2 + 6x - 50$	$y' = 4x + 6$
$y = 2x^2 + 6x + \sqrt{10}$	$y' = 4x + 6$
$y = 2x^2 + 6x + c$	$y' = 4x + 6$

where c is any constant.

The opposite operation (that is, going from the right-hand column to the left-hand column) is called *integration*. The answer to an integration problem is called the *indefinite integral*. As is the case with factoring, integration does not yield a single (unique) answer.

Example 5   Integration

Derivative of a Function	Function (Indefinite Integral)
$y' = 4x + 6$	$y = 2x^2 + 6x + c$

The letter c represents any constant, and c is called the *constant of integration*.

The original constant is generally not known when integrating. Therefore the constant c is used to represent all possible constants. Hence the indefinite integral is not unique. (In some applied problems the constant can be determined.)

DEFINITION

Let $y = f(x)$. The function $F(x) + c$ is called the indefinite integral of $y = f(x)$ provided

$$y = f(x) = \frac{d}{dx}[F(x) + c].$$

Example 6   The indefinite integral is not unique, but it is a family of functions with similar graphs. Suppose $y' = 2x$. Then $y = x^2 + c$. (See Figure 5.1.) All curves in the family have the same derivative at $x = p$.

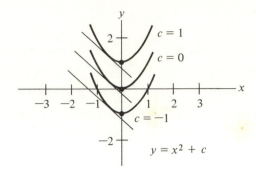

FIGURE 5.1

Example 7

Function	Indefinite Integral	Reason
$y = 2x + 3$	$F(x) + c = x^2 + 3x + c$	$\dfrac{d}{dx}(x^2 + 3x + c) = 2x + 3$
$g(t) = e^t$	$G(t) + c = e^t + c$	$\dfrac{d}{dt}(e^t + c) = e^t$
$f(x) = 3x^2 + 5x$	$F(x) + c = x^3 + \dfrac{5x^2}{2} + c$	$\dfrac{d}{dx}\left(x^3 + \dfrac{5x^2}{2} + c\right) = 3x^2 + 5x$

As with other mathematical operations, a symbol is used to indicate when the operation is to be performed. The symbol $\int dx$ is used to indicate integration. The symbol $\int$ is read "the integral of" and dx indicates that the independent variable is x. ($\int dz$ means to find the indefinite integral where the independent variable is z.) Using this notation, we can symbolize Example 7.

Example 8 (Example 7 Symbolized)

$$\int (2x + 3)\, dx = x^2 + 3x + c$$
$$\int e^t\, dt = e^t + c$$
$$\int (3x^2 + 5x)\, dx = x^3 + \frac{5x^2}{2} + c$$

5.2 Formulas for Integration

There are several formulas that are useful for the operation of integration.

FORMULA 1

$$\int 0\, dx = c.$$

The indefinite integral of zero is a constant.

Proof

$$\frac{d}{dx} c = 0$$

Example 9

1. $\int 0\, dx = c$
2. $\int 0\, dz = c$
3. $\int 0\, dw = c$

FORMULA 2

$\int k\, dx = kx + c$, where k and c are constants.

The indefinite integral of a constant is the constant multiplied by the independent variable plus c.

Proof

$$\frac{d}{dx}(kx + c) = k(1) + 0 = k$$

Example 10

1. $\int 6\, dx = 6x + c$
2. $\int 6\, dz = 6z + c$
3. $\int 6\, dw = 6w + c$

Note the need for the dx, dz, or dw part of the integration symbol to indicate the independent variable.

The concept of a constant multiplier in an integration problem is so important that a separate formula is needed to express it.

FORMULA 3

$\int k\, f(x)\, dx = k \int f(x)\, dx$, where k is a constant.

The indefinite integral of a constant k times a function is the constant times the integral of the function.

Example 11

1. $\int 5f(x)\, dx = 5 \int f(x)\, dx$
2. $\int 3x^{1/2}\, dx = 3 \int x^{1/2}\, dx$
3. $\int 2e^t\, dt = 2 \int e^t\, dt$

FORMULA 4

$$\int x^n \, dx = \frac{x^{n+1}}{n+1} + c, \text{ where } n \neq -1.$$

The indefinite integral of x to a constant exponent is x to the exponent plus 1, all divided by the exponent plus 1.

Example 12

1. $\int x^3 \, dx = \dfrac{x^4}{4} + c$

2. $\int x^{1/2} \, dx = \dfrac{2x^{3/2}}{3} + c$

3. $\int n^{-2} \, dn = -n^{-1} + c$

By using Formulas 3 and 4, we can integrate the following:

4. $\int 5x^3 \, dx = 5 \int x^3 \, dx = 5\left(\dfrac{x^4}{4}\right) + c$

5. $\int 7x^5 \, dx = 7 \int x^5 \, dx = 7\left(\dfrac{x^6}{6}\right) + c$

FORMULA 5

If $y = f(x) + g(x) + \cdots + h(x)$, then

$$\int y \, dx = \int f(x) \, dx + \int g(x) \, dx + \cdots + \int h(x) \, dx.$$

If a function consists of a finite number of terms, the indefinite integral is found by adding together the integrals of each term.

Example 13

1. $\int (4x^2 + x - 3) \, dx = \int 4x^2 \, dx + \int x \, dx - \int 3 \, dx$

$$= \frac{4x^3}{3} + \frac{x^2}{2} - 3x + c$$

2. $\int (2n - n^2) \, dn = \int 2n \, dn - \int n^2 \, dn$

$$= n^2 - \frac{n^3}{3} + c$$

3. $\int (2x^{-5} + 1) \, dx = \dfrac{2x^{-4}}{-4} + x + c$

$$= \frac{-1}{2x^4} + x + c$$

Formula 6 is the opposite of Rule 5 for differentiation.

FORMULA 6

$$\int [f(x)]^n \cdot f'(x)\, dx = \frac{[f(x)]^{n+1}}{n+1} + c, \text{ where } n \neq -1.$$

Example 14   Find $\int (x^2 + 3)^6(2x)\, dx$.

Now, $f(x) = x^2 + 3$; therefore $f'(x) = 2x$. Hence $(x^2 + 3)^6(2x)$ is of the form $[f(x)]^n \cdot f'(x)$. Applying Formula 6, we have

$$\int (x^2 + 3)^6(2x)\, dx = \frac{(x^2 + 3)^7}{7} + c.$$

Example 15   Find $\int (2x^3 - 5)^4 x^2\, dx$.

Let $f(x) = 2x^3 - 5$; then $f'(x) = 6x^2$. Therefore the expression $(2x^3 - 5)^4(x^2)$ is not of the form $[f(x)]^n \cdot f'(x)$, and $(2x^3 - 5)^4(x^2) \neq (2x^3 - 5)^4(6x^2)$, precluding use of Formula 6. However, we can supply the necessary 6 and $\frac{1}{6}$, yielding $6(\frac{1}{6}) \int (2x^3 - 5)^4 x^2\, dx$. Applying Formula 3 and Formula 6 yields

$$\tfrac{1}{6} \int (2x^3 - 5)^4(6x^2)\, dx = \frac{(2x^3 - 5)^5}{30} + c$$

Example 16   Find $\int 3x(x^2 - 4)^7\, dx$.

In this example, $f(x) = x^2 - 4$; therefore $f'(x) = 2x$. Again, the expression $3x(x^2 - 4)^7$ is not in the form $[f(x)]^n \cdot f'(x)$. Rewriting the problem by Formula 3 yields $3 \int x(x^2 - 4)^7\, dx$. We can supply the necessary 2 and $\frac{1}{2}$; hence

$$3(\tfrac{1}{2}) \int 2x(x^2 - 4)^7\, dx = \frac{3(x^2 - 4)^8}{16} + c$$

Note: Do not overestimate the power of this method. Variables, even when their product is 1, cannot be inserted on opposite sides of the integral symbol. This can be done with *constants only.*

Nonexample 17   Find $\int (x^3 - 3)^{1/2}\, dx$.

Let $f(x) = x^3 - 3$; then $f'(x) = 3x^2$. Since the factor x^2 is not contained in the function to be integrated, Formula 6 does not apply. Therefore we cannot solve $\int (x^3 - 3)^{1/2}\, dx$ by Formula 6.

Example 18   Find $\int (x^2 - 6x - 8)^{1/2}(2x - 6)\, dx$.

Since $f(x) = x^2 - 6x - 8$, $f'(x) = 2x - 6$. Therefore Formula 6 applies, and we have

$$\int (x^2 - 6x - 8)^{1/2}(2x - 6)\, dx = \frac{2(x^2 - 6x - 8)^{3/2}}{3} + c$$

▶ **Exercise 5.2**

Find the indefinite integral in Problems 1 through 25.

1 $\int 2\,dx$ 2 $\int -5\,dw$

3 $\int (x + 3)\,dx$ 4 $\int (2x^2 + 3x - 5)\,dx$

5 $\int (3 - 2x + x^4)\,dx$ 6 $\int \left(6x^2 - \dfrac{x}{2}\right) dx$

7 $\int x^{1/2}\,dx$ 8 $\int 2\sqrt{x}\,dx$

9 $\int x\,dx$ 10 $\int (y^2 - 6y)\,dy$

11 $\int (w^{-2} + w^{1/3})\,dw$ 12 $\int (3x + a)\,dx$, where a is a constant

13 $\int (x^2 + bx)\,dx$, where b is a constant 14 $\int (5x^3 - 5)^4 x^2\,dx$

15 $\int (2 - 3x^2)x\,dx$ 16 $\int (x^2 + 3x^3)^4(2x + 9x^2)\,dx$

17 $\int (x^2 + 3)x\,dx$ 18 $\int (x^2 - 6)^{1/3}x\,dx$

19 $\int (2x^3 + 1)^{99}x^2\,dx$ 20 $\int (5x^3 + 1)^6 x^2\,dx$

21 $\int (3x^4 - 10)^{-2}x^3\,dx$ 22 $\int (3x + 4)^{1/2}\,dx$

23 $\int (9x^2 + 2)^{1/3}x\,dx$ 24 $\int (2y + 3)\,dy$

25 $\int \dfrac{x^2 + 2x^3}{x}(1 + 4x)\,dx$ (*Hint:* Reduce the fraction.)

26 Prove Formula 4.

5.3 Integration of $e^{g(x)}$

Formula 7 is the opposite of Rule 8 for differentiation.

FORMULA 7

$$\int g'(x)e^{g(x)}\,dx = e^{g(x)} + c.$$

The indefinite integral of $e^{g(x)}$ multiplied by $g'(x)$ is $e^{g(x)}$ plus c.

Proof

$$\frac{d}{dx}\left(e^{g(x)} + c\right) = g'(x)e^{g(x)}$$

Example 19 Find $\int 6e^{6x}\,dx$.

Since $g(x) = 6x$, $g'(x) = 6$. Therefore $6e^{6x}$ is of the form $g'(x)\,e^{g(x)}$. Applying Formula 7, we have

$$\int 6e^{6x}\,dx = e^{6x} + c$$

Example 20 Find $\int 3x^2 e^{x^3}\,dx$.

Since $g(x) = x^3$, $g'(x) = 3x^2$. Applying Formula 7, we have

$$\int 3x^2 e^{x^3}\,dx = e^{x^3} + c$$

Example 21 Find $\int 2e^{6x}\, dx$.

In this example $g(x) = 6x$; therefore $g'(x) = 6$. However the function $2e^{6x} \neq 6e^{6x}$, precluding use of Formula 7. By Formula 3, we may rewrite the problem as $2 \int e^{6x}\, dx$. Supplying 6 and $\frac{1}{6}$, we have

$$\tfrac{2}{6} \int 6e^{6x}\, dx = \tfrac{1}{3}e^{6x} + c$$

Applied Example 22 The rate of radioactive disintegration of an element can be represented by the equation $dn/dt = n_0(-k)e^{-kt}$, where dn/dt represents the rate of atoms disintegrating with respect to time t. The constant n_0 is the number of radioactive atoms present at time $t = 0$, and k is the disintegration constant of the particular element under consideration.

1. Find the integral of dn/dt. The constant of integration is zero. (Why?)

$$n = \int \left(\frac{dn}{dt}\right) dt = \int n_0(-k)e^{-kt}\, dt = n_0 e^{-kt}$$

The equation $n = n_0 e^{-kt}$ represents the number n of atoms that are still radioactive after time t.

The half-life of a radioactive element is the time when one half of the original number of radioactive atoms have disintegrated. (Thus half the original number are still radioactive.) The number of radioactive atoms at the time when the half-life is attained can be represented by $n = n_0/2$.

2. Assume for some element that $n_0 = 10^{30}$ and $k = .01$ per day. Find the half-life of the element in days.

Since $n = n_0 e^{-kt}$ and $n = n_0/2$, where $n_0 = 10^{30}$, we have $10^{30}/2 = 10^{30}e^{-.01t}$. Dividing by 10^{30}, we have $\frac{1}{2} = e^{-.01t}$. Let $x = .01t$; then $.5 = e^{-x}$, and using the table in Appendix 3, we find that $x = .7$. Therefore $.01t = .7$, and $t = 70$ days.

3. Graph the function $n = n_0 e^{-.01t}$. (See Figure 5.2.)

FIGURE 5.2

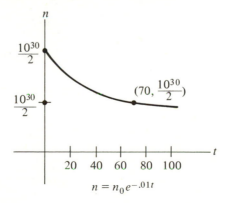

$$n = n_0 e^{-.01t}$$

▶ **Exercise 5.3**

Find the indefinite integral in Problems 1 through 11.

1 $\int e^x \, dx$ 2 $\int 3e^{3x} \, dx$

3 $\int 2xe^{x^2} \, dx$ 4 $\int 12x^3 e^{3x^4} \, dx$

5 $\int (2x - 2)e^{x^2 - 2x} \, dx$ 6 $\int e^x \cdot e^{e^x} \, dx$

7 $\int 2e^x \, dx$ 8 $\int xe^{5x^2} \, dx$

9 $\int xe^{-x^2} \, dx$ 10 $\int e^{-x} \, dx$

11 $\int 10e^{-.2t} \, dt$

12 Assume that $n_0 = 10^{20}$ and $k = .02$ per day.
   a Find the half-life of the element.
   b Find the rate of change of n with respect to t at $t = 10$, at $t = 30$.

5.4 Integration by Parts

The derivative of a product of two functions is determined by Rule 6. If $y = f(x) g(x)$, then $y' = f(x) g'(x) + f'(x) g(x)$. Integrating, $y = \int f(x) g'(x) \, dx + \int g(x) f'(x) \, dx$. Solving for $\int f(x) g'(x) \, dx$ leads to Formula 8.

FORMULA 8

If the function to be integrated is of the form $f(x) g'(x)$, then

$$\int f(x) g'(x) \, dx = f(x) g(x) - \int f'(x) g(x) \, dx.$$

Use the following procedure for applying Formula 8.

1 Decide which of the two functions in the product to be integrated is to be $f(x)$ and which is to be $g'(x)$. Usually, $f(x)$ is chosen to be the function whose derivative is simpler than the function itself.

2 Find $f'(x)$ and $g(x)$, where $g(x)$ is an integral of $g'(x)$.

3 Apply Formula 8.

Example 23 Find $\int xe^x \, dx$.
   Let $f(x) = x$; then $f'(x) = 1$. Let $g'(x) = e^x$; then $g(x) = e^x$. Applying Formula 8,

$$\int xe^x \, dx = xe^x - \int e^x \, dx = xe^x - e^x + c$$

Example 24 Find $\int x(x - 3)^4 \, dx$.
   Let $f(x) = x$; then $f'(x) = 1$. Let $g'(x) = (x - 3)^4$; then $g(x) = (x - 3)^5/5$. Applying Formula 8,

$$\int x(x - 3)^4 \, dx = \frac{x(x - 3)^5}{5} - \int \frac{(x - 3)^5}{5} \, dx = \frac{x(x - 3)^5}{5} - \frac{(x - 3)^6}{30} + c$$

Example 25 Find $\int x/\sqrt{x+3}\, dx$.

Let $f(x) = x$; then $f'(x) = 1$. Let $g'(x) = (x+3)^{-1/2}$; then $g(x) = (x+3)^{1/2}/\frac{1}{2} = 2(x+3)^{1/2}$. By Formula 8,

$$\int \frac{x}{\sqrt{x+3}}\, dx = 2x(x+3)^{1/2} - \int 2(x+3)^{1/2}\, dx$$

$$= 2x(x+3)^{1/2} - \frac{2(x+3)^{3/2}}{\frac{3}{2}} + c$$

$$= 2x\sqrt{x+3} - \frac{4\sqrt{(x+3)^3}}{3} + c$$

▶ **Exercise 5.4**

Find the indefinite integral in each of the following.

1 $\int 2xe^x\, dx$

2 $\int \frac{xe^x}{5}\, dx$

3 $\int x(x+5)^6\, dx$

4 $\int 2x(x-1)^5\, dx$

5 $\int x(x+2)^{-1/2}\, dx$

6 $\int \frac{3x}{(3x+1)^{1/2}}\, dx$

7 $\int x(2x+7)^{1/3}\, dx$

8 $\int x^2 e^x\, dx$

5.5 Summary and Comments on Integration

In this chapter we have developed several integration formulas. They are listed here for reference.

Formula 1 $\int 0\, dx = c$.

Formula 2 $\int k\, dx = kx + c$, where k is any constant.

Formula 3 $\int k \cdot f(x)\, dx = k \int f(x)\, dx$, where k is any constant.

Formula 4 $\int x^n\, dx = \dfrac{x^{n+1}}{n+1} + c$, where $n \neq -1$.

Formula 5 $\int [f(x) + g(x) + \cdots + h(x)]\, dx$
$$= \int f(x)\, dx + \int g(x)\, dx + \cdots + \int h(x)\, dx.$$

Formula 6 $\int [f(x)]^n \cdot f'(x)\, dx = \dfrac{[f(x)]^{n+1}}{n+1} + c$, where $n \neq -1$.

Formula 7 $\int g'(x)\, e^{g(x)}\, dx = e^{g(x)} + c$.

Formula 8 $\int f(x)\, g'(x)\, dx = f(x)\, g(x) - \int f'(x)\, g(x)\, dx$.

There are other integration formulas. Most good mathematical handbooks give long lists of them. Once the concept of integration is mastered, the more extensive tables can be used in finding the integral.

▶ **Exercise 5.5**

Find the indefinite integrals.

1 $\int (x^3 + 3e^{3x})\, dx$ 2 $\int (3x^2 + 3e^{3x})(x^3 + e^{3x})^2\, dx$

3 $\int (x + x^{-2})\, dx$ 4 $\int (e^x + 2)\, dx$

5 $\int (e^x + e)\, dx$ 6 $\int x^e\, dx$

7 $\displaystyle\int \frac{dx}{(3x + 4)^{1/2}}$ 8 $\int -3e^x\, dx$

9 $\int x^2(6x^3 + e)^{-3}\, dx$ 10 $\displaystyle\int \frac{1}{e^x}\, dx$

11 $\displaystyle\int \frac{x\, dx}{(x^2 - 5)^{1/2}}$ 12 $\int e^e\, dx$

5.6 Application of the Indefinite Integral

The importance of marginal functions was shown in Chapter 3. If we know the marginal function, we can integrate and find the total function plus a constant. In applied problems the value of the constant usually can be determined.

Applied Example 26 Let $C'(x) = 2 + x$ be a marginal cost function. The total cost is $C = \int (2 + x)\, dx = 2x + x^2/2 + c$, where x is the number of units produced. If no units are produced, then $C(0) = 2(0) + 0^2/2 + c = c$. Therefore the constant of integration is the fixed cost.

Applied Example 27 Let $R'(x) = 2x - x^2$ be a marginal revenue function. The total revenue function is $R(x) = x^2 - x^3/3 + c$, where x is the number of units sold. It is reasonable to assume that if no units are sold, there is no revenue. Therefore the constant of integration is zero and $R(x) = x^2 - x^3/3$.

A profit function is slightly more complicated. Profit is revenue minus cost, $P(x) = R(x) - C(x)$, where x is the number of units produced and sold.

Applied Example 28 Let marginal revenue $R'(x) = 5$ and marginal cost $C'(x) = x/8$. Hence $R(x) = 5x$ and $C(x) = x^2/16 + c$, where c is the fixed cost. Profit is $P(x) = 5x - x^2/16 - c$.

Applied Example 29 A firm's retail sales depend on the amount t spent on advertising. If $S'(t) = 10e^{-.01t}$ is the marginal sales function, find the total sales function.

$$S(t) = \int 10e^{-.01t}\, dt = -1000e^{-.01t} + c$$

Further, the firm knows that it will have sales of 1000 with no advertising. Therefore at $t = 0$,

$$S(0) = -1000e^{-.01(0)} + c = 1000$$
$$-1000(1) + c = 1000$$
$$c = 2000$$

The total sales function is $S(t) = 2000 - 1000e^{-.01t}$. If $t = 100$, then $S(100) = 2000 - 1000e^{-1} = 2000 - 368 = 1632$.

Example 30 If the derivative of a function is $y' = 4x - 4$, find the indefinite integral.

$$y = \int (4x - 4)\, dx = 2x^2 - 4x + c$$

Find the member of the family that passes through the point (2, 1). At this point,

$$2(2)^2 - 4(2) + c = 1$$
$$0 + c = 1$$
$$c = 1$$

The desired member is $y = 2x^2 - 4x + 1$.

▶ Exercise 5.6

1 If the marginal cost is $C'(x) = 50x + 1000$, find the total cost function. If the fixed cost is 2000, find the cost of producing 50 units.

2 An object is projected upward with a velocity $v = 320 - 32t$, where t is the time elapsed after the object is projected upward. Find the altitude function. At $t = 0$, the distance traveled is zero. Find the time t where the velocity is zero. Find the maximum height of the object.

3 If the marginal profit is $P'(x) = 10 - 2x$, find the total profit function. If $c = -10$, what level of sales will yield the maximum profit? What is the maximum profit?

4 If the derivative of a function is $f'(x) = 2x + 3$, find the indefinite integral. Which member of the family passes through the point (0, −4)?

5 A firm's retail sales depend on the amount spent on advertising. The marginal sales function is $s' = 5e^{-.02t}$. Find the total sales function if the firm's sales are 500 with no advertising.

6 If the marginal revenue is $R'(x) = 100 - 2x$, find the total revenue function. What is the revenue when $x = 50$ units?

7 If the derivative of a function is $f'(x) = 2x^2 - 6x$, find the indefinite integral that passes through the point (1, 6).

8 If the marginal output is $Q'(x) = 2x + 3$, find the total output function. If the constant of integration is zero, what is the output when the input $x = 50$?

9 If the marginal cost is $C'(x) = 4x + 500$, find the total cost function when the fixed cost is 500.

10 If the marginal profit is $P'(x) = 200 - 10x$, find the total profit function when the fixed cost is 500. What is the maximum profit?

11 Let the marginal revenue be $R'(x) = 150$. Let the marginal cost be $C'(x) = 2x + 17$.
a Find the total revenue function.
b Find the total cost function if the fixed cost is 1000.
c Find the break-even points, that is, where the cost is equal to the revenue.
d Write the profit equation.
e What is the profit (loss) when production is zero?
f At what level of production is the profit a maximum?
g If the company is to make *any* profit, what range of production must be maintained?

12 To determine the desired average weight of his women patients, a doctor uses the fact that the rate of increase in weight per inch of increase in height is 23/7 (for 25-year-olds of medium frame). Find the equation that relates weight as a function of height if a woman 5 ft 0 in. tall weighs 100 lb. How much should a woman who is 5 ft 4 in. weigh?

13 A firm estimates that its rate of growth in sales R for the next 20 years will be $R' = .1t^{1/4}$. If the firm's present sales are 1 (in millions of dollars per year), find the total sales function. What will the firm's sales be in 16 years?

14 The value of an item decreases with time by the rate $v'(t) = -50e^{-t}$. Find the value function if the original value was 50. What is the value of the item after 3 years?

15 Given its present facilities, a company has determined that the rate of change of units produced P with respect to an increase in workers x is $P' = x^{-1/2}$. Find the production equation if present production is 100 units. What increase in production can be expected by the addition of 16 workers?

16 The marginal increase in heartbeat with respect to an increase of 1° in body temperature is 10. If the normal heartbeat is 60 beats/min at a temperature of 98° F, express heartbeat as a function of temperature (over the domain of 98° to 105°). What is the heartbeat if the temperature is 100°?

5.7   Differential Equations

An equation that contains a derivative is called a *differential equation*. In this section we will consider equations that have only one independent variable, since the derivative implies one independent variable.

Example 31   Each of the following is a differential equation.

$$1. \quad \frac{dy}{dx} = 2x - 3, \text{ independent variable } x$$

$$2. \quad \frac{dy}{dx} = 2xe^{x^2}, \text{ independent variable } x$$

$$3. \quad \frac{dy}{dt} = \frac{-1}{t^2}, \text{ independent variable } t$$

A solution to a differential equation of the form $dy/dx = f(x)$ is a function $y = F(x)$, which has the property that $F'(x) = f(x)$.

Example 32   Solutions to the differential equations in Example 31 are as follows:

$$1. \quad y = x^2 - 3x + c$$
$$2. \quad y = e^{x^2} + c$$
$$3. \quad y = \frac{1}{t} + c$$

The differential equations in this discussion contain only differentials of first order because every derivative is to the first power. Thus far we have considered the symbol dy/dx as an entity. However, if the symbol is considered as the quotient of two quantities, dy and dx, we can write $dy/dx = f(x)$ as $dy = f(x)\,dx$. The quantities dy and dx are called *differentials*, and we say that the equation $dy = f(x)\,dx$ is in its differential form.

Example 33

Derivative Notation	Differential Notation
$1. \quad \dfrac{dy}{dx} = 2x - 3$	$dy = (2x - 3)\,dx$
$2. \quad \dfrac{dy}{dx} = 2xe^{x^2}$	$dy = 2xe^{x^2}\,dx$
$3. \quad \dfrac{dy}{dt} = \dfrac{-1}{t^2}$	$dy = \dfrac{-1}{t^2}\,dt$

Differential notation facilitates separation of the variable y and its differential dy from the variable x and its differential dx. Differential equations in which the variables can be separated are called *separable differential equations*. Unfortunately, not all differential equations are separable.

Example 34

Differential Equation	Variables Separated
1. $\dfrac{dy}{x} = \dfrac{dx}{y}$	$y\,dy = x\,dx$
2. $dy = y^2 e^x\,dx$	$\dfrac{dy}{y^2} = e^x\,dx$
3. $(2x - y)\,dx + (3x + 2y)\,dy = 5$	Not separable

If the variables can be separated as in Example 34, parts 1 and 2, an equation without differentials can be found by integrating both sides of the separated differential equation. The equation obtained is called a *solution of the first-order differential equation.*

Example 35 Solutions to Example 34, parts 1 and 2, follow.

1. $\int y\,dy = \int x\,dx$ 2. $\int \dfrac{dy}{y^2} = \int e^x\,dx$

$\dfrac{y^2}{2} + c_1 = \dfrac{x^2}{2} + c_2$ $\dfrac{-1}{y} + c_1 = e^x + c_2$

Combining the constants,   Combining the constants,

$\dfrac{y^2}{2} - \dfrac{x^2}{2} = c$ $e^x + \dfrac{1}{y} = c$

Example 36 Solve $(x/e^y)\,dx = dy$.

Separating the variables,

$$x\,dx = e^y\,dy$$

Integrating,

$$\int x\,dx = \int e^y\,dy$$
$$\dfrac{x^2}{2} + c_1 = e^y + c_2$$

Combining the constants,

$$\dfrac{x^2}{2} - e^y = c$$

Applied Example 37 The rate of change of pressure on one square foot of the ocean floor due to a change of depth is represented by the differential equation $dp/dh = 64.5$. Assume that $p = 0$ when $h = 0$. Determine the pressure when the depth is 100.

Separating the variables,

$$dp = 64.5\,dh$$

Integrating,

$$\int dp = \int 64.5\,dh$$

Therefore $p = 64.5h + c$. Since $p = 0$ when $h = 0$, then $c = 0$. Thus

$$p = 64.5h$$
$$p(100) = 64.5(100) = 6450$$

▶ **Exercise 5.7**

Solve the differential equations in Problems 1 through 7.

1 $\dfrac{dR}{dP} = .3$

2 $\dfrac{dy}{dx} = 3x - 4$

3 $\dfrac{dy}{dx} = 2e^{2x}$

4 $\dfrac{dy}{dx} = \dfrac{x}{y^2}$

5 $\dfrac{dy}{dx} = 4x^3$

6 $\dfrac{dy}{dx} = 2x + \dfrac{1}{x^2}$

7 $\dfrac{dy}{dx} = xy^3$

8 The rate of change of velocity of a moving object is 20 ft/sec². Find the velocity as a function of time, assuming that $v = 0$ when $t = 0$. Find the velocity at $t = 1$ sec.

9 In a circuit with constant resistance, the rate of change of voltage dv/dt is directly proportional to the rate of change of current dI/dt. Find the equation relating voltage to current. Assume $v = 0$ when $I = 0$. Find the voltage when the current is .7.

---

self-test • chapter five

Find the indefinite integral in Problems 1 through 8.

1 $\int \sqrt{x}\, dx$

2 $\int (3x^2 + 5x - 6)\, dx$

3 $\int \dfrac{3}{x^2}\, dx$

4 $\int x\sqrt{x^2 + 3}\, dx$

5 $\int \dfrac{2}{\sqrt{3x - 4}}\, dx$

6 $\int xe^{3x^2+1}\, dx$

7 $\int (2 + 3e^{-x})\, dx$

8 $\int xe^{-x}\, dx$

9 If the marginal cost function is given by $C'(x) = 3x + 5$, find the cost function. Assume that the fixed cost is $1000.

10 Solve the differential equation $dy/dx = x - 5$.

11 Solve the differential equation $dy = (2x/y^3)\, dx$.

12 A firm's marginal sales function is $S' = 5e^{-.02t}$, where t is the amount of money spent on advertising. Find the sales function if the firm sells 100 units with no advertising.

chapter six

the definite integral

6.1 The Fundamental Theorem of Calculus

In Chapter 5 we developed techniques for finding indefinite integrals. Now we will combine the indefinite integral with functional evaluations in order to obtain a real number. This process is called *finding the definite integral*. The method of combining the indefinite integral with certain functional evaluations is given in the following rule, commonly called the *Fundamental Theorem of Calculus* (hereafter called the F.T.C.).

FUNDAMENTAL THEOREM OF CALCULUS (F.T.C.)

Let $y = f(x)$ be a continuous function over the closed interval from $x = a$ to $x = b$. That is, $a \leq x \leq b$. Let $F(x)$ be an indefinite integral of $y = f(x)$. Then the definite integral is a number given by

$$\int_a^b f(x)\, dx = F(x) \Big|_a^b = F(b) - F(a)$$

Example 1 Evaluate $\displaystyle\int_2^3 (x^2 + 2)\, dx$.

$$\int_2^3 (x^2 + 2)\, dx = \left(\frac{x^3}{3} + 2x\right)\Big|_2^3$$

$$F(x) = \frac{x^3}{3} + 2x$$

$$F(b) = F(3) = \frac{3^3}{3} + 2(3) = 9 + 6 = 15$$

$$F(a) = F(2) = \frac{2^3}{3} + 2(2) = \frac{8}{3} + 4 = \frac{20}{3}$$

$$F(b) - F(a) = F(3) - F(2) = 15 - \frac{20}{3} = \frac{25}{3}$$

Thus $25/3$ is the value of the definite integral when $f(x) = x^2 + 2$, $a = 2$, and $b = 3$. (Note that $F(x)$ does not contain the number c, since it is subtracted out in determining $F(b) - F(a)$.)

Example 2 Evaluate $\displaystyle\int_{-1}^{2} (x - 2)\, dx$.

$$\int_{-1}^{2} (x - 2)\, dx = \left(\frac{x^2}{2} - 2x \right)\Bigg|_{-1}^{2}$$

$$F(x) = \frac{x^2}{2} - 2x$$

$$F(b) = F(2) = -2$$

$$F(a) = F(-1) = \frac{5}{2}$$

$$F(b) - F(a) = F(2) - F(-1) = \frac{-9}{2}$$

Therefore $-9/2$ is the value of the definite integral when $f(x) = x - 2$, $a = -1$, and $b = 2$.

Example 3 Evaluate $\displaystyle\int_{0}^{2} (x^3 - x)\, dx$.

$$\int_{0}^{2} (x^3 - x)\, dx = \left(\frac{x^4}{4} - \frac{x^2}{2} \right)\Bigg|_{0}^{2}$$

$$F(x) = \frac{x^4}{4} - \frac{x^2}{2}$$

$$F(b) = F(2) = 2$$

$$F(a) = F(0) = 0$$

$$F(b) - F(a) = F(2) - F(0) = 2$$

▶ **Exercise 6.1**

Find the value of the definite integral in each of the following problems.

1 $\displaystyle\int_{-1}^{1} x^2\, dx$ **2** $\displaystyle\int_{0}^{2} e^x\, dx$

3 $\displaystyle\int_{1}^{4} \sqrt{x}\, dx$ **4** $\displaystyle\int_{2}^{4} (2 - x^2)\, dx$

5 $\displaystyle\int_1^5 \frac{1}{x^2}\,dx$

6 $\displaystyle\int_{-2}^2 (x^3 - 1)\,dx$

7 $\displaystyle\int_{-1}^2 (x^3 + 4x^2 - 2x + 1)\,dx$

8 $\displaystyle\int_0^1 xe^{x^2}\,dx$

9 $\displaystyle\int_2^4 x\sqrt{x^2 - 4}\,dx$

6.2   The Definite Integral Related to Area

In every example using the F.T.C. a specific real number was found. But what does this real number represent? Consider the following:

1 Let $f(x) \geq 0$ for all x such that $a \leq x \leq b$.

2 Partition the interval a, b into n equal intervals. Each interval is of length $(b - a)/n$. (See Figure 6.1.)

3 Select a value x_i from each interval and compute $f(x_i)$. The value x_i may be one of the endpoints of the interval. In what follows, we will select x_i as an endpoint.

4 Compute the areas of each rectangle formed and then add these areas together.

$$A = \frac{b - a}{8} \cdot f(a) + \frac{b - a}{8} \cdot f(x_1) + \frac{b - a}{8} \cdot f(x_3) + \frac{b - a}{8} \cdot f(x_4)$$

$$+ \frac{b - a}{8} \cdot f(x_5) + \frac{b - a}{8} \cdot f(x_6) + \frac{b - a}{8} \cdot f(x_6) + \frac{b - a}{8} \cdot f(x_7)$$

This area is less than the area bounded by $y = f(x)$, the vertical lines $x = a$, $x = b$, and the x axis. However, it is an approximation of the area. A better approximation could be found by taking more intervals.

FIGURE 6.1

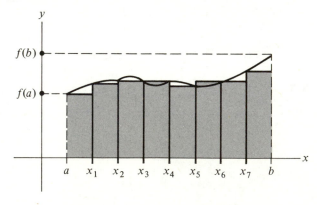

The exact area of this bounded region could be found by allowing the number of intervals to increase without bound and then taking the limit of the sum. Symbolically, this is written as follows:

$$A = \lim_{n \to \infty} \sum_{i=1}^{n} \frac{b - a}{n} \cdot f(x_i)$$

where $f(x_i)$ is a value from step 3. The Σ is the notation for summation. For example, A could have been written as

$$A = \sum_{i=0}^{7} f(x_i) \frac{b - a}{8}$$

where we let $x_0 = a$.

The task of finding the limit can be difficult. Although some of the techniques of Chapter 2 could be used in finding the limit, we will not try to evaluate the limit by those earlier techniques. Instead we will make the following definition and use this definition to compute the limits.

DEFINITION

$$A = \lim_{n \to \infty} \sum_{i=1}^{n} \frac{b - a}{n} \cdot f(x_i) = \int_{a}^{b} f(x)\, dx = F(b) - F(a)$$

provided $y = f(x)$ is continuous and $f(x) \geq 0$ on the closed interval from $x = a$ to $x = b$. Also, $F(x)$ is an indefinite integral of $f(x)$.

This definition means that the definite integral is an area when $f(x) \geq 0$ for all x in the closed interval from a to b. If this seems too much to believe, consider the following examples.

FIGURE 6.2

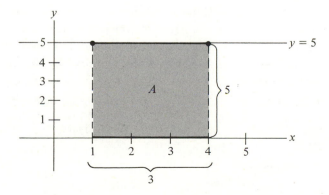

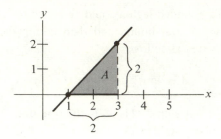

FIGURE 6.3

Example 4   Find the area bounded by the constant function $y = 5$, the lines $x = 1$, $x = 4$, and the x axis. Note that the region is a rectangle. (See Figure 6.2.)

Area by geometry	*Area by using the definition*

$A = lw$ (l length, w width)

$A = (3)(5)$

$A = 15$ sq units

$$A = \int_1^4 5\,dx = 5x \Big|_1^4 = 5(4) - 5(1)$$

$$= 20 - 5$$

$$= 15$$

Therefore the area by geometry agrees with the area obtained by the definition.

Example 5   Find the area bounded by the function $y = x - 1$, the lines $x = 1$, $x = 3$, and the x axis. Note that the region is a triangle. (See Figure 6.3.)

Area by geometry   *Area by using the definition*

$A = \frac{1}{2}bh$ (b base, h height)

$A = \frac{1}{2}(2)(2)$

$A = 2$ sq units

$$A = \int_1^3 (x - 1)\,dx = \frac{x^2}{2} - x \Big|_1^3$$

$$= 1\tfrac{1}{2} - (-\tfrac{1}{2})$$

$$= 2$$

Therefore the area found by geometry and the area found by the definition are the same when the region is a triangle.

Example 6   Find the area bounded by the function $y = 3x + 1$, the lines $x = 1$, $x = 3$, and the x axis. Note that the region is a trapezoid. (See Figure 6.4.)

A trapezoid is a four-sided plane figure in which two sides are parallel. The two parallel sides are called *bases*, and the shortest distance between the bases is called the *altitude*. The area of a trapezoid can be found by the formula $A = \frac{1}{2}h(b_1 + b_2)$, where b_1 and b_2 are the two bases and h is the altitude.

Area by geometry   *Area by using the definition*

$A = \frac{1}{2}h(b_1 + b_2)$

$A = \frac{1}{2}(2)(4 + 10)$

$A = 14$ sq units

$$A = \int_1^3 (3x + 1)\,dx = \frac{3x^2}{2} + x \Big|_1^3$$

$$= 16\tfrac{1}{2} - 2\tfrac{1}{2}$$

$$= 14$$

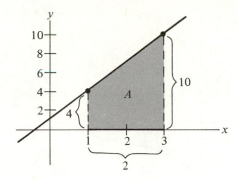

FIGURE 6.4

Again we obtain the same value for the area. Hence we conclude that our definition does actually produce an area. We also conclude that the value of the definite integral is the area of a bounded region provided that $f(x) \geq 0$ for all values of x in the closed interval from $x = a$ to $x = b$.

Example 7 Find the area bounded by the function $y = x^3 + 8$, the lines $x = 2$, $x = -1$, and the x axis. (See Figure 6.5.)

Area by geometry

There is no formula in geometry that will find such an area.

Area by using the definition

$$A = \int_{-1}^{2} (x^3 + 8)\, dx = \left. \frac{x^4}{4} + 8x \right|_{-1}^{2}$$

$$= 20 - (-7\tfrac{3}{4})$$

$$= 27\tfrac{3}{4}$$

FIGURE 6.5

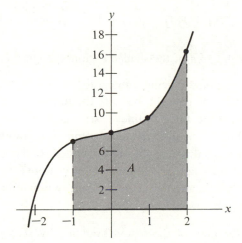

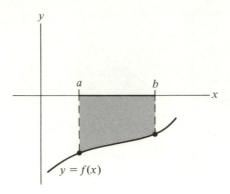

y = f(x)

FIGURE 6.6

This example illustrates a limitation of the use of geometry in finding area. Calculus prevails!

Example 8   Find the area bounded by $f(x) = x^2 + 3$, the vertical lines $x = 1$, $x = 3$, and the x axis. (See Figure 6.6.)

$$\int_a^b f(x)\,dx = \int_1^3 (x^2 + 3)\,dx = \left(\frac{x^3}{3} + 3x\right)\Bigg|_1^3$$

$$F(b) = F(3) = \frac{3^3}{3} + 3(3) = 18$$

$$F(a) = F(1) = \frac{1^3}{3} + 3(1) = \frac{10}{3}$$

$$F(b) - F(a) = F(3) - F(1) = 18 - \frac{10}{3} = \frac{44}{3} = 14\frac{2}{3}$$

Thus $14\frac{2}{3}$ sq units is the area of the shaded region in Figure 6.6.

To make general use of the definite integral for finding area, we must give further consideration to the concept, because the function need not be positive over an interval. There are three distinct cases that must be considered. (See Figures 6.7, 6.8, and 6.9.) The first, the most common in applied problems, has already been discussed. In this case the area bounded by the graph of the function, the two vertical lines, and the x axis is always above the x axis, and the F.T.C. is used directly to compute the area of the region. But what if the area of the region is always below the x axis?

Example 9   Find the area of the region bounded by $f(x) = x^3$, the vertical lines $x = -2$, $x = 0$, and the x axis. (See Figure 6.10.)

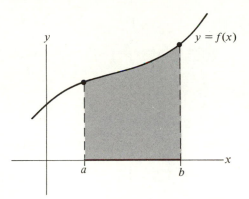

FIGURE 6.7

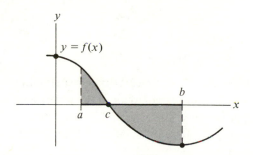

FIGURE 6.8

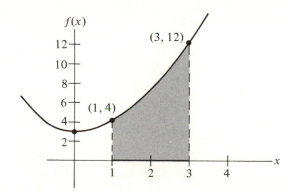

FIGURE 6.9

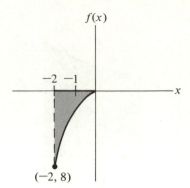

f(*x*)

−2 −1

x

FIGURE 6.10

(−2, 8)

By the F.T.C.,

$$\int_{-2}^{0} x^3 \, dx = \frac{x^4}{4} \Big|_{-2}^{0}$$

$$F(0) = 0$$

$$F(-2) = 4$$

$$F(0) - F(-2) = -4$$

The value of the integral is −4, but clearly the area should always be positive. Consequently the area is considered to be 4 sq units, and the negative sign indicates that the area is below the *x* axis. The area of a region that is never above the *x* axis is the absolute value of the definite integral. The area for Example 9 is 4 sq units.

Example 10 Find the area of the region bounded by $f(x) = -2 - x^2$, the vertical lines $x = -1$, $x = 1$, and the *x* axis (See Figure 6.11.)

FIGURE 6.11

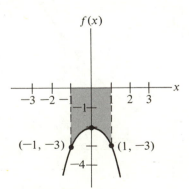

f(*x*)

−3 −2 −1   1 2 3   *x*

−1

(−1, −3)   (1, −3)

−4

By the F.T.C.,

$$\int_{-1}^{1}(-2 - x^2)\, dx = \left(-2x - \frac{x^3}{3}\right)\Bigg|_{-1}^{1}$$

$$F(1) = -2\tfrac{1}{3}$$

$$F(-1) = 2\tfrac{1}{3}$$

$$F(1) - F(-1) = -4\tfrac{2}{3}$$

Since the area of the desired region is below the x axis, the area is the absolute value of $-4\tfrac{2}{3}$, which is $4\tfrac{2}{3}$ sq units.

There is still one remaining possibility. Some of the area bounded by the function, the two vertical lines, and the x axis may be above the x axis, while the rest of the area may be below the x axis. In this case the point where the curve intersects the x axis (x intercept) must be found. The area of the region above the x axis must be computed separately from the area of the region below the x axis. To find the total area the separate areas must be added.

Example 11   Find the area of the region bounded by $f(x) = x^3$, the vertical lines $x = -2$, $x = 1$, and the x axis.

From Figure 6.12 it can be seen that the curve crosses the x axis at $x = 0$. The region below the x axis is from $x = -2$ to $x = 0$, and the region above the x axis is from $x = 0$ to $x = 1$.

To complete this problem, we must note another property of integration, which indicates that the total area is the sum of two areas. It is stated as follows:

$$\int_{a}^{b} f(x)\, dx = \int_{a}^{c} f(x)\, dx + \int_{c}^{b} f(x)\, dx$$

FIGURE 6.12

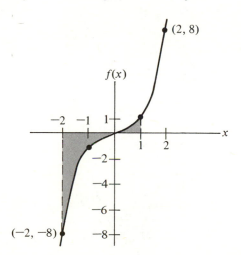

where $a < c < b$.

$$\int_{-2}^{1} x^3\, dx = \int_{-2}^{0} x^3\, dx + \int_{0}^{1} x^3\, dx$$

since $-2 < 0 < 1$. Then

$$\int_{-2}^{0} x^3\, dx = \frac{x^4}{4} \Big|_{-2}^{0}$$

$$F(0) = 0$$
$$F(-2) = 4$$
$$F(0) - F(-2) = -4$$

Thus the area of the region below the x axis is 4 sq units.

$$\int_{0}^{1} x^3\, dx = \frac{x^4}{4} \Big|_{0}^{1}$$

$$F(1) = \tfrac{1}{4}$$
$$F(0) = 0$$
$$F(1) - F(0) = \tfrac{1}{4}$$

Therefore the area of the region above the x axis is $\tfrac{1}{4}$ sq units. The total area is $4 + \tfrac{1}{4} = 4\tfrac{1}{4}$ sq units. (*Note:* To find the value of x where the curve crosses the x axis, set the function equal to zero and solve the resulting equation for x.)

Example 12   Find the area of the region bounded by $f(x) = x^2 - 7x + 10$, the vertical lines $x = 0$, $x = 4$, and the x axis.

Figure 6.13 indicates that 2 is an x intercept.

FIGURE 6.13

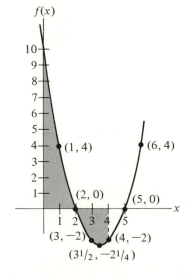

To be sure, set $x^2 - 7x + 10 = 0$. Then $(x - 5)(x - 2) = 0$, and $x = 5$ or $x = 2$. However, since $x = 2$ is between $x = 0$ and $x = 4$, the area must be computed in two parts.

$$\int_0^2 (x^2 - 7x + 10)\, dx = \frac{x^3}{3} - \frac{7x^2}{2} + 10x \Big|_0^2$$

$$F(2) = 8\tfrac{2}{3}$$

$$F(0) = 0$$

$$F(2) - F(0) = 8\tfrac{2}{3}$$

Thus the area of the region above the x axis is $8\tfrac{2}{3}$ sq units.

$$\int_2^4 (x^2 - 7x + 10)\, dx = \frac{x^3}{3} - \frac{7x^2}{2} + 10x \Big|_2^4$$

$$F(4) = \frac{16}{3}$$

$$F(2) = \frac{26}{3}$$

$$F(4) - F(2) = \frac{-10}{3} = -3\tfrac{1}{3}$$

Thus the area of the region below the x axis is $3\tfrac{1}{3}$ sq units. The total area of the region is found by adding the two individual areas.

$$\text{Area} = 8\tfrac{2}{3} + 3\tfrac{1}{3} = 12 \text{ sq units}$$

▶ **Exercise 6.2**

In Problems 1 through 15 find the areas bounded by the functions, the lines $x = a$, $x = b$, and the x axis.

1 $y = 5$, $x = 1$ and $x = 3$

2 $y = 2x - 4$, $x = 2$ and $x = 5$

3 $y = x - 2$, $x = 2$ and $x = 4$

4 $y = 2x + 1$, $x = 0$ and $x = 3$

5 $y = 3 - x$, $x = 0$ and $x = 3$

6 $y = x + 4$, $x = 1$ and $x = 3$

7 $y = x^3$, $x = 1$ and $x = 2$

8 $y = x^2$, $x = 2$ and $x = 4$

9 $y = \sqrt{x}$, $x = 1$ and $x = 4$

10 $f(x) = 3x^2 - 2x + 4$; $x = 2$ and $x = 3$

11 $f(x) = 4x^3 - 1$; $x = 1$ and $x = 3$

12 $f(x) = e^x$; $x = 0$ and $x = 2$

13 $f(x) = x + e^x$; $x = 0$ and $x = 1$

14 $f(x) = 2xe^{x^2}$; $x = 0$ and $x = 1$

15 $f(x) = 1/x^2$; $x = 1$ and $x = 2$

16 Find the area of the region bounded by $f(x) = -x^2 + 4$ and the x axis. (*Hint:* Find the x intercepts.)

17 Find the area of the region bounded by $f(x) = -x^2 + 2x + 3$ and the x axis.

In Problems 18 through 23 find the areas bounded by the functions, the two vertical lines, and the x axis. In each problem it should be determined that the desired region is never above the x axis. There are two methods of determining the location of the region with respect to the x axis. First, the function can be graphed and the region observed. Second, $f(a)$ and $f(b)$ can be computed. If both $f(a)$ and $f(b)$ are negative and there is no x intercept between a and b, the entire region is below the x axis.

18 $f(x) = -x^2$; $x = -1$ and $x = 1$

19 $f(x) = 6x^2 + 3x - 4$; $x = -1$ and $x = 0$

20 $f(x) = 1/x^3$; $x = -2$ and $x = -1$

21 $y = -e^{-x}$; $x = 0$ and $x = 1$

22 $y = 3x - 6$; $x = 0$ and $x = 2$

23 $y = x - 3$; $x = 1$ and $x = 2$

In Problems 24 through 29 find the areas bounded by the functions, the two vertical lines, and the x axis. (Check each function for an x intercept between the vertical lines.)

24 $f(x) = x - 1$, $x = 0$ and $x = 2$

25 $f(x) = 3x^2 - 3$, $x = 0$ and $x = 2$

26 $f(x) = x^2 - 3x + 2$, $x = 1$ and $x = 3$

27 $f(x) = x^3 + 8$, $x = -3$ and $x = -1$

28 $f(x) = x - 2$, $x = 1$ and $x = 3$

29 $f(x) = (x - 1)(x - 2)(x - 3)$, $x = 0$ and $x = 4$

6.3   The Area Between Two Curves

There are times when it is necessary to discuss areas of regions that do not have the horizontal axis as one boundary. The region may be enclosed by two functions that intersect at two distinct points, or it may be bounded by two functions and two vertical lines. We consider these cases separately.

In Figure 6.14 it is evident that the area of the bounded region is actually the area of the region bounded by the function $y = f(x)$, the horizontal axis, and the two vertical lines $x = a$ and $x = b$ minus the area of the region bounded by the function $y = g(x)$, the horizontal axis, and the vertical lines $x = a$ and $x = b$.

Since the area of each of the individual regions can be found by the application of the F.T.C., the area of the region bounded by the two functions can be found by subtraction, that is, by

$$\int_a^b f(x)\, dx - \int_a^b g(x)\, dx$$

The difference of the two integrals can also be written as

$$\int_a^b [f(x) - g(x)]\, dx$$

Either form of the difference of the two integrals may be used to compute the desired area. The first form is a bit longer, but it has the advantage of eliminating possible errors of computation.

Example 13   Find the area of the region bounded by $g(x) = x^2$ and $f(x) = -x^2 + 8$. (See Figure 6.15.)

Since the vertical lines are missing from the description of the area, the values of a and b must be determined. Both functions have the same functional value at

FIGURE 6.14

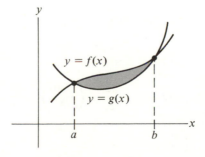

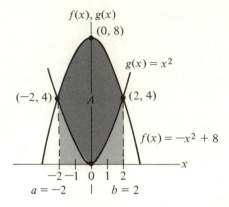

FIGURE 6.15

the points of intersection. The values of a and b can be determined by setting the two functions equal to each other and solving the resulting equation for x.

$$x^2 = -x^2 + 8$$

$$2x^2 = 8$$

$$x^2 = 4$$

$$x = 2 \quad \text{or} \quad x = -2$$

$$a = -2 \quad \text{and} \quad b = 2$$

Now the integral can be used to compute the desired area A, where A_1 is the area beneath $f(x) = -x^2 + 8$, and A_2 is the area beneath $g(x) = x^2$.

$$A = A_1 - A_2 = \int_{-2}^{2} (-x^2 + 8)\, dx - \int_{-2}^{2} x^2\, dx$$

By working with the two individual areas, the following result is obtained.

$$A_1 = \int_{-2}^{2} (-x^2 + 8)\, dx = \frac{-x^3}{3} + 8x \Big|_{-2}^{2}$$

$$F(2) = \frac{40}{3}$$

$$F(-2) = \frac{-40}{3}$$

$$F(2) - F(-2) = \frac{40}{3} - \left(\frac{-40}{3}\right) = \frac{80}{3}$$

Thus $A_1 = 80/3$.

$$A_2 = \int_{-2}^{2} x^2\, dx = \frac{x^3}{3}\Big|_{-2}^{2}$$

$$G(2) = \frac{8}{3}$$

$$G(-2) = \frac{-8}{3}$$

$$G(2) - G(-2) = \frac{8}{3} - \left(\frac{-8}{3}\right) = \frac{16}{3}$$

Thus $A_2 = 16/3$.

$$A = A_1 - A_2 = \frac{80}{3} - \frac{16}{3} = \frac{64}{3}\ \text{sq units}$$

Example 14 Find the area bounded by $y = x^2 - 2x + 3$ and $y = -x^2 - x + 6$. (See Figure 6.16.)

First the values of a and b must be determined. Set the two equations equal to each other.

$$x^2 - 2x + 3 = -x^2 - x + 6$$

$$2x^2 - x - 3 = 0$$

$$(2x - 3)(x + 1) = 0$$

$$x = \tfrac{3}{2} \quad \text{or} \quad x = -1$$

$$a = -1 \quad \text{and} \quad b = \tfrac{3}{2}$$

FIGURE 6.16

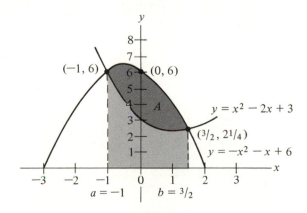

Then

$$A_1 = \int_{-1}^{3/2} (-x^2 - x + 6)\,dx = \left. \frac{-x^3}{3} - \frac{x^2}{2} + 6x \right|_{-1}^{3/2}$$

$$F(\tfrac{3}{2}) = \frac{27}{4}$$

$$F(-1) = -\frac{37}{6}$$

$$A_1 = F(\tfrac{3}{2}) - F(-1) = \frac{27}{4} - \left(\frac{-37}{6}\right) = \frac{155}{12}$$

Then

$$A_2 = \int_{-1}^{3/2} (x^2 - 2x + 3)\,dx = \left. \frac{x^3}{3} - x^2 + 3x \right|_{-1}^{3/2}$$

$$F(\tfrac{3}{2}) = \frac{27}{8}$$

$$F(-1) = -\frac{13}{3}$$

$$A_2 = F(\tfrac{3}{2}) - F(-1) = \frac{27}{8} - \left(-\frac{13}{3}\right) = \frac{185}{24}$$

Therefore

$$A = A_1 - A_2 = \frac{155}{12} - \frac{185}{24} = \frac{125}{24} = 5\tfrac{5}{24} \text{ sq units}$$

It may also be the case that the region is bounded by two functions and two vertical lines $x = a$ and $x = b$. (See Figure 6.17.) The same procedure is

FIGURE 6.17

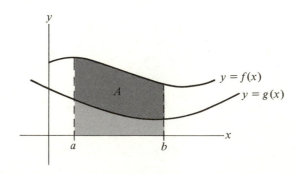

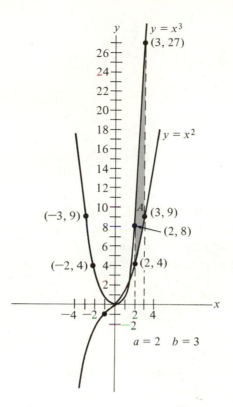

FIGURE 6.18

followed in this type of problem with one exception: The values of a and b must be given.

Example 15 Find the area of the region bounded by $y = x^3$, $y = x^2$ and the vertical lines $x = 2$, $x = 3$. (See Figure 6.18.)

$$A_1 = \int_2^3 x^3\, dx = \frac{x^4}{4}\Big|_2^3$$

$$F(3) = \frac{81}{4}$$

$$F(2) = 4$$

$$F(3) - F(2) = \frac{81}{4} - 4 = \frac{65}{4}$$

Also,

$$A_2 = \int_2^3 x^2\, dx = \frac{x^3}{3}\Big|_2^3$$

$$F(3) = 9$$

$$F(2) = \frac{8}{3}$$

$$F(3) - F(2) = 9 - \frac{8}{3} = \frac{19}{3}$$

Therefore

$$A = A_1 - A_2 = \frac{65}{4} - \frac{19}{3} = \frac{119}{12} \text{ sq units}$$

In the method used for finding the areas individually, the problem becomes more difficult when part of the desired region is below the horizontal axis. Since there are two distinct functions given in such a way that one of the functions is always "above" the other in the closed interval, $f(x) - g(x)$ is always positive and the F.T.C. will work directly in computing the area. In this case it is convenient to use

$$\int_a^b [f(x) - g(x)]\, dx$$

rather than

$$\int_a^b f(x)\, dx - \int_a^b g(x)\, dx$$

Example 16   Find the area bounded by $f(x) = 2x - 5$ and $g(x) = x^2 - 4x$. (See Figure 6.19.)

FIGURE 6.19

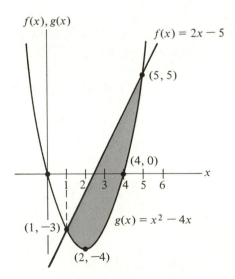

First we must find the values of a and b.

$$x^2 - 4x = 2x - 5$$
$$x^2 - 6x + 5 = 0$$
$$(x - 5)(x - 1) = 0$$
$$x = 5 \quad \text{or} \quad x = 1$$
$$a = 1 \quad \text{and} \quad b = 5$$

Then

$$\int_1^5 (2x - 5)\, dx - \int_1^5 (x^2 - 4x)\, dx = \int_1^5 [(2x - 5) - (x^2 - 4x)]\, dx$$

$$= \int_1^5 (2x - 5 - x^2 + 4x)\, dx$$

$$= \int_1^5 (-x^2 + 6x - 5)\, dx$$

$$= \left. \frac{-x^3}{3} + 3x^2 - 5x \right|_1^5$$

$$F(5) = \frac{25}{3}$$

$$F(1) = -\frac{7}{3}$$

$$F(5) - F(1) = \frac{32}{3}$$

Therefore $A = 32/3$ sq units.

▶ Exercise 6.3

1 Find the area of the region bounded by $y = x^2$ and $y = x$.

2 Find the area of the region bounded by $y = x^2 - 6x + 10$ and $y = x$.

3 Find the area of the region bounded by the curve $f(x) = \sqrt{x}$ and $g(x) = 1/x^2$ and the vertical lines $x = 1$, $x = 4$.

4 Find the area of the region bounded by $f(x) = e^x$, $g(x) = 8$, and the vertical lines $x = 0$, $x = 2$.

5 Find the area of the region bounded by $f(x) = -x^2 + 9$, $g(x) = 5$, and the y axis.

6 Find the area of the region bounded by $f(x) = 2x - 7$ and $g(x) = x^2 - 6x$.

6.4 Applications of the Definite Integral

One of the most useful applications of the definite integral is in the field of probability. Determining the probability of success or failure is of utmost importance in decision making. Even though it is possible to study probability without an understanding of calculus, the insight gained from calculus makes many probability ideas clearer.

In a finite sample space the probability of an event occurring is a number between 0 and 1. Zero probability means that the event can never occur. A probability of 1 means that the event must always occur. A probability of .63 means that the favorable outcome should occur 63 times out of every 100 outcomes.

In order to use the definite integral as a tool in probability, we must define a probability density function. Such a function will be denoted by $y = P(x)$, and it must satisfy the following properties:

1 $y = P(x)$ must be continuous for all x such that $a \leq x \leq b$.

2 $P(x) \geq 0$ for all x such that $a \leq x \leq b$.

3 $\displaystyle\int_a^b P(x)\,dx = 1.$

Example 17 Verify that $P(x) = 2x$ is a probability density function when $a = 0$ and $b = 1$.

Since $P(x) = 2x$ is a polynomial function, it is continuous for all x such that $0 \leq x \leq 1$. Therefore property 1 is satisfied.

We have $P(x) \geq 0$ for all x such that $0 \leq x \leq 1$. Therefore property 2 is satisfied.

By the F.T.C.

$$\int_0^1 2x\,dx = 2\int_0^1 x\,dx = x^2\,\Big|_0^1 = (1)^2 - (0)^2 = 1$$

Therefore property 3 is satisfied. Since all three properties are satisfied, $P(x) = 2x$ is a probability density function on the interval $[0, 1]$.

If $y = P(x)$ is a probability density function, the probability that a particular event will occur over a desired interval, say c to d, in the fixed interval a to b, is given by the following:

4 $\displaystyle\int_c^d P(x)\,dx.$

Example 18 Find the probability of an event occurring between $x = \frac{1}{2}$ and $x = \frac{3}{4}$ when $P(x) = 2x$ and the fixed interval is $a = 0$ and $b = 1$.

$P(x) = 2x$ is a probability density function from $a = 0$ to $b = 1$, since it satisfies the three properties. (See Example 17.) Thus the probability is

$$\int_{1/2}^{3/4} 2x\, dx = x^2 \Big|_{1/2}^{3/4} = (\tfrac{3}{4})^2 - (\tfrac{1}{2})^2 = .3125$$

If a situation occurs in which we have a constant times a function, it is possible to find the constant so that the function will be a probability density function.

Example 19 Let $P(x) = k(x^2 + 1)$. Find the value of k that will make the function a probability density function for the interval from $a = 0$ to $b = 1$.

Consider $f(x) = x^2 + 1$. Since this is a polynomial function, it is continuous on the interval from $a = 0$ to $b = 1$. Since $f(x) = x^2 + 1$ and the domain of x is $0 \le x \le 1$, $f(x) \ge 0$ for all x in the domain. By the F.T.C.

$$\int_0^1 (x^2 + 1)\, dx = \frac{x^3}{3} + x \Big|_0^1 = \frac{4}{3} - 0 = \frac{4}{3}$$

Since $4/3 \ne 1$, $f(x) = x^2 + 1$ is not a probability density function. Now $3/4$ times $4/3 = 1$. Thus a constant multiplier of $k = 3/4$ will make $P(x) = (3/4)(x^2 + 1)$ a probability density function. Thus $P(x) = (3/4)(x^2 + 1)$ satisfies all three properties. Verify that

$$\int_0^1 (3/4)(x^2 + 1)\, dx = 1.$$

Generalizing, k can be computed by the following formula:

$$k = \frac{1}{\displaystyle\int_a^b f(x)\, dx}$$

Applied Example 20 If $P(x) = (3/4)(x^2 + 1)$ describes the frequency of filling x orders in a one-week period, what is the probability that 80% or better of the orders will be filled in 1 week? (Let $a = 0$ and $b = 1$, since the percentage can vary from 0% to 100%.)

$P(x) = (3/4)(x^2 + 1)$ is a probability density function. This was shown in Example 19. Thus the desired probability is

$$\frac{3}{4} \int_{.8}^1 (x^2 + 1)\, dx = \frac{3}{4} \left(\frac{x^3}{3} + x \right) \Big|_{.8}^1 = 1.000 - .728 = .272$$

Therefore the probability of filling 80% or more of the orders is .272 (or 27.2%).

Another application of the definite integral is with respect to marginal functions, in particular the marginal output function. Given a marginal output function that is always a positive continuous function in an interval, the bounded area is the increase in total output.

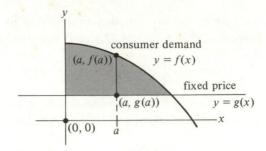

FIGURE 6.20

Applied Example 21 Let $y' = x^3 + x + 1$ be a marginal output function. Find the change in total output when the input changes from $x = 20$ to $x = 40$.

$$\int_{20}^{40} (x^3 + x + 1)\, dx = \frac{x^4}{4} + \frac{x^2}{2} + x \Big|_{20}^{40} = F(40) - F(20)$$

$$F(40) = 640{,}840$$

$$F(20) = 40{,}220$$

$$F(40) - F(20) = 640{,}840 - 40{,}220 = 600{,}620$$

Thus 600,620 is the increase in total output y when the input changes from $x = 20$ to $x = 40$.

The next two topics illustrate practical uses for the area between two curves. The first deals with the economics principle called consumer surplus. Consider the difference between the price that is being charged for a product and the price that the consumer is willing to pay for that product. If the consumer is willing to pay more than the price being charged, he has gained financially by paying the market price. This gain is shown in Figure 6.20 by $f(a) - g(a)$ at $x = a$, where x is the number of items produced. The area shown in the figure, which is the sum of all of the strips $f(a) - g(a)$ from $x = 0$ to $x = b$, is called the consumer surplus.

Applied Example 22 Let the consumer demand function be given by $f(x) = -x^2 + 6$, where x is the number of items produced. The price is a constant given by $g(x) = 2$. Find the consumer surplus. (See Figure 6.21.) Since the function is restricted by the application, $a = 0$. We must find the value of b by determining where the two graphs intersect. This can be done by setting the two equations equal to each other and solving for x. Thus $-x^2 + 6 = 2$. Solving for x, we have $x = 2$ or $x = -2$. The solution $x = -2$ can be discarded, since it is not an admissible answer. Thus $b = 2$. The consumer surplus is found by

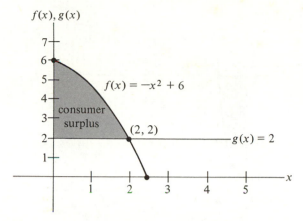

FIGURE 6.21

$$\int_0^2 (-x^2 + 6)\, dx - 2 \int_0^2 dx = \frac{-x^3}{3} + 6x - 2x \Big|_0^2 = F(2) - F(0)$$

$$F(2) = \frac{16}{3}$$

$$F(0) = 0$$

$$F(2) - F(0) = \frac{16}{3} - 0 = \frac{16}{3}$$

Another application of the area between two curves is in the topic of producer's surplus. Consider the difference between the fixed market price of an item and the price at which the producer is willing to sell the item. If the market price is greater than the price for which the producer would sell the item, then the producer has increased his revenue by selling at the fixed market price. This increase in revenue is noted in Figure 6.22 by $g(a) - f(a)$. The area shown in

FIGURE 6.22

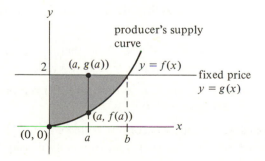

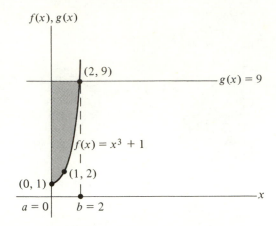

FIGURE 6.23

the figure, which is the sum of all the strips $g(a) - f(a)$ from $x = 0$ to $x = b$, is called the producer's surplus.

Applied Example 23   Let $f(x) = x^3 + 1$ be the producer's supply function, where x is the number of units produced. Let $g(x) = 9$ be the fixed market price. Find the producer's surplus. (See Figure 6.23.)

Now, $a = 0$ because of the applied restriction. Solving for b, we have $x^3 + 1 = 9$. Then $x^3 = 8$ and $x = 2$. Thus $b = 2$.

$$\int_0^2 9 \, dx - \int_0^2 (x^3 + 1) \, dx = 9x - \frac{x^4}{4} - x \Big|_0^2 = F(2) - F(0)$$

$$F(2) = 12$$
$$F(0) = 0$$
$$F(2) - F(0) = 12 - 0 = 12$$

▶ **Exercise 6.4**

1 Show that $f(x) = 1$ is a probability density function on the interval from $x = 2$ to $x = 3$.

2 Let $f(x) = k(x + 3)$. Find the value of k so that $y = f(x)$ is a probability density function on the interval from $x = 1$ to $x = 2$.

3 Let $f(x) = ke^{-x}$. Find the value of k so that $y = f(x)$ is a probability density function on the interval from $x = 0$ to $x = 1$.

4 A company fills its weekly order according to the equation $f(x) = k(2 + x^2)$. Find the value of k so that $y = f(x)$ is a probability density function for the interval from $x = 0$ to $x = 1$. Find the probability that the company will fill 90% or better of its orders for the week.

5 A bottle factory has determined that its marginal output function per hour is given by $y' = x^3 + 400$, where x is the number of assembly lines in operation. What is the increase in the hourly production of bottles when the number of assembly lines is increased from five to ten?

6 A store manager determined that his marginal profit function was given by $P' = x^{-2/3}$, where x is the number of hours that his salesmen work in a day. Find the increase in profit when the manager requires his salesman to work 10 hr/day instead of 8 hr/day.

7 Let $f(x) = -x^2 + 2x + 4$ be the consumer demand function. Let $g(x) = 4$ be the fixed price. Find the consumer surplus.

8 Let $f(x) = -x^2 + 10$ be the consumer demand function. Let $g(x) = 9$ be the fixed price. Find the consumer surplus.

9 Let $f(x) = x^2$ be the producer's supply function. Let $g(x) = 4$ be the fixed marked price. Find the producer's surplus.

6.5   Improper Integrals

The values determined by the conditions in integrals such as

$$\textbf{1} \ \int_1^\infty \frac{1}{x^2}\,dx \qquad \textbf{2} \ \int_{-\infty}^\infty e^{-x^2}\,dx \qquad \textbf{3} \ \int_0^1 \frac{1}{1 - x^2}\,dx$$

$$\textbf{4} \ \int_0^3 \frac{1}{x^3}\,dx \qquad \textbf{5} \ \int_{-1}^2 \frac{1}{x^2}\,dx$$

cannot be computed by using the F.T.C. They do not satisfy the conditions of the F.T.C. for several reasons. In **1**, b is not finite. In **2**, neither a nor b is finite. Integral **3** is discontinuous at $x = 1$, and integral **4** is discontinuous at $x = 0$. Integral **5** is discontinuous at $x = 0$, which is between the bounds of integration, which are $x = -1$ and $x = 2$. Integrals such as these are called *improper integrals*.

Since the values of the above integrals cannot be directly computed by $F(b) - F(a)$, the following question seems appropriate: Can a value be found when the bounds of integration are not finite or where the function is not continuous on the closed interval from $x = a$ to $x = b$? The only answer that can be given to that question is, sometimes yes, sometimes no. Therefore we must establish a means of answering this question about integrals with a definite yes or a definite no.

There are three basic types of improper integrals. The first type is one that allows either one or both of the bounds of integration to be infinity. (See integrals **1** and **2**.) This cannot be done by the F.T.C. using $F(b) - F(a)$, since a and b must be finite in that theorem. (There must be a closed interval.)

Example 24   In $\int_1^\infty 1/x^2\, dx$ (integral **1**), the F.T.C. could be used for large values of b. Note the results obtained in Table 6.1. As $b \to \infty$, $F(b) - F(a) \to 1$.

TABLE 6.1

Value of b	Integral	F(x)	F(b)	F(a)	F(b) − F(a)
100	$\int_1^{100} 1/x^2\, dx$	$-1/x$	$-1/100$	-1	.99
1000	$\int_1^{1000} 1/x^2\, dx$	$-1/x$	$-1/1000$	-1	.999
1,000,000	$\int_1^{1,000,000} 1/x^2\, dx$	$-1/x$	$-1/1,000,000$	-1	.999999

Therefore the $\int_1^\infty 1/x^2\, dx$ exists, and the area of the region bounded by this function, the x axis, and the vertical line $x = 1$ is 1, because the integral $\int_1^b 1/x^2\, dx$ converges to 1 as b increases without bound. (See Figure 6.24.)

Symbolically, we write this result in the following notation:

$$\int_1^\infty \frac{1}{x^2}\, dx = \lim_{b \to \infty} \int_1^b \frac{1}{x^2}\, dx = \lim_{b \to \infty} \frac{-1}{x}\Big|_1^b$$

$$F(b) = \frac{-1}{b} \quad \text{and} \quad F(a) = -1$$

$$\lim_{b \to \infty} [F(b) - F(a)] = \lim_{b \to \infty}\left[\frac{-1}{b} - (-1)\right]$$

$$= \lim_{b \to \infty} \frac{-1}{b} - \lim_{b \to \infty}(-1) = 0 + 1 = 1$$

The second basic type of improper integral may be recognized by the fact that at least one of the values of a and b is a value for which the function is discontinuous. (See integral **4**.)

Example 25   The function $f(x) = 1/\sqrt{x}$ is not continuous at $x = 0$, since $f(0) = 1/0$, which is undefined. Thus $\int_0^4 1/\sqrt{x}\, dx$ cannot be found by the F.T.C., because the conditions for using the F.T.C. require that $f(x) = 1/\sqrt{x}$

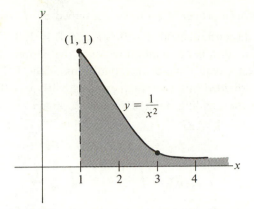

FIGURE 6.24

be continuous on $0 \le x \le 4$. However, the F.T.C. can be used for values of x that are larger than zero and that approach zero. Note results of Table 6.2.

TABLE 6.2

Value of a	Integral	F(x)	F(b)	F(a)	F(b) − F(a)
1	$\int_{1}^{4} 1/\sqrt{x}\,dx$	$2\sqrt{x}$	4	2	2
1/4	$\int_{1/4}^{4} 1/\sqrt{x}\,dx$	$2\sqrt{x}$	4	1	3
1/100	$\int_{1/100}^{4} 1/\sqrt{x}\,dx$	$2\sqrt{x}$	4	1/5	3.8
1/1,000,000	$\int_{1/1,000,000}^{4} 1/\sqrt{x}\,dx$	$2\sqrt{x}$	4	1/500	3.998

As $a \to 0$ from the right, $F(b) - F(a) \to 4$. Thus it can be stated that the area of the region identified by $\int_{0}^{4} 1/\sqrt{x}\,dx$ is finite (or exists) and that the area is 4.

Symbolically, we write the following.

$$\int_{0}^{4} \frac{1}{\sqrt{x}}\,dx = \lim_{a \to 0} \int_{a}^{4} \frac{1}{\sqrt{x}}\,dx = \lim_{a \to 0} (2\sqrt{x})\Big|_{a}^{4}$$

$$\lim_{a \to 0} [F(b)] = \lim_{a \to 0} (2\sqrt{4}) = 4$$

$$\lim_{a \to 0} [F(a)] = \lim_{a \to 0} (2\sqrt{a}) = 0$$

$$\lim_{a \to 0} [F(b) - F(a)] = \lim_{a \to 0} F(b) - \lim_{a \to 0} F(a) = 4 - 0 = 4$$

The third type of improper integral is exemplified by $\int_{-1}^{2} 1/x^2\, dx$ (integral **5**). This function is discontinuous at $x = 0$, since $f(0) = 1/0$. Since 0 is between -1 and 2, this function is not continuous over the entire interval from $x = a$ to $x = b$. This fact violates one of the conditions of the F.T.C. Therefore this problem must be divided into two distinct parts, with $x = 0$, the point of discontinuity, being the dividing point. Example 26 analyzes the behavior of the improper integral $\int_{-1}^{2} 1/x^2\, dx$.

Example 26

$$\int_{-1}^{2} \frac{1}{x^2}\, dx = \int_{-1}^{0} \frac{1}{x^2}\, dx + \int_{0}^{2} \frac{1}{x^2}\, dx$$

In $\int_{0}^{2} 1/x^2\, dx$, the F.T.C. will apply if zero is approached from the positive side by using such values as $1/2$, $1/10$, $1/100$, and $1/1{,}000{,}000$ for a. (See Table 6.3.)

TABLE 6.3

Value of a	Integral	F(x)	F(b)	F(a)	F(b) − F(a)
1/2	$\int_{1/2}^{2} 1/x^2\, dx$	$-1/x$	$-1/2$	-2	1.5
1/10	$\int_{1/10}^{2} 1/x^2\, dx$	$-1/x$	$-1/2$	-10	9.5
1/100	$\int_{1/100}^{2} 1/x^2\, dx$	$-1/x$	$-1/2$	-100	99.5
1/1,000,000	$\int_{1/1,000,000}^{2} 1/x^2\, dx$	$-1/x$	$-1/2$	$-1{,}000{,}000$	999,999.5

As $a \to 0$ from the right, $F(b) - F(a) \to \infty$. Since this limit is not finite, the improper integral does not exist. It is said to diverge. It is not necessary to compute the other integral in this problem, because the divergence of one of the two parts is sufficient to force the original integral to diverge.

$$\int_{-1}^{2} \frac{1}{x^2}\, dx = \lim_{b \to 0} \int_{-1}^{b} \frac{1}{x^2}\, dx + \lim_{a \to 0} \int_{a}^{2} \frac{1}{x^2}\, dx$$

Then

$$\lim_{a \to 0} \int_{0}^{2} \frac{1}{x^2}\, dx = \lim_{a \to 0} \frac{-1}{x}\Big|_{a}^{2}$$

$$\lim_{a \to 0} F(b) = \lim_{a \to 0} \frac{-1}{2} = \frac{-1}{2}$$

$$\lim_{a \to 0} F(a) = \lim_{a \to 0} \frac{-1}{a} = -\infty$$

Therefore $\lim_{a\to0} [F(b) - F(a)]$ is undefined. If this integral had converged, we would have had to evaluate $\lim_{b\to0} \int_{-1}^{b} 1/x^2\, dx$ before the final decision on convergence or divergence could have been made.

It should be noted that any improper integral must either converge or diverge. The three types of improper integrals discussed in the previous examples simply outline the method of attack. Once an integral is identified as an improper integral of a specific type, the question of convergence depends on the evaluation of the proper limit. That limit must be finite in order for the integral to converge.

▶ **Exercise 6.5**

Determine the convergence or divergence of each of the improper integrals in Problems 1 through 6. If it converges, find the value to which it converges.

1 $\displaystyle\int_{1}^{\infty} \frac{1}{x^3}\, dx$
2 $\displaystyle\int_{1}^{\infty} \frac{1}{\sqrt{x}}\, dx$

3 $\displaystyle\int_{0}^{\infty} e^{-x}\, dx$
4 $\displaystyle\int_{2}^{6} \frac{1}{\sqrt{x-2}}\, dx$

5 $\displaystyle\int_{-1}^{1} \frac{1}{x^4}\, dx$
6 $\displaystyle\int_{0}^{1} \frac{1}{\sqrt[3]{x}}\, dx$

7 The formula for finding the present value of a perpetual income stream flowing at a uniform rate of D dollars/year and at a continuous discount rate r is given by

$$P = \int_{0}^{\infty} De^{-rt}\, dt,$$

where t is time.

a What is the present value of a perpetual cash flow of $2000/year when $r = 7\%$?

b What is the present value of a perpetual cash flow of $1500/year when $r = 6\%$?

c Show that

$$P = \int_{0}^{\infty} De^{-rt}\, dt$$

always reduces to $P = D/r$.

6.6 Approximation Method for Finding Area

Since the advent of electronic computers, many arithmetic techniques that require long, tedious calculations have been revived. This section is devoted to finding the area bounded by the lines $x = a$, $x = b$, the x axis, and the given

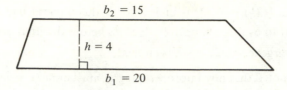

$b_2 = 15$

$h = 4$

$b_1 = 20$

FIGURE 6.25

function $y = f(x)$, provided that $f(x)$ is continuous and not negative for all values of x between a and b. This particular technique for approximating the area is based on an area formula found in plane geometry.

Example 27 Find the area of the trapezoid whose bases are 15 and 20 units respectively and whose altitude is 4 units. (See Figure 6.25.)

$$A = \tfrac{1}{2}(4)(20 + 15) = 2(35) = 70 \text{ sq units}$$

With this area formula as background, let us consider the area of the region bounded by $f(x) = x^2$, the x axis, and the two vertical lines $x = 1$, $x = 3$. (See Figure 6.26.)

Partition the interval $1 \leq x \leq 3$ into four equal parts. This gives the x values of 1, 3/2, 2, 5/2, 3. Find $f(x)$ for each of those x values, and connect the points $(x, f(x))$ in order to form four trapezoids. The bases of these trapezoids are two vertical lines whose lengths are equal to $f(x)$ for each specific value of x. The altitude is the distance between each two consecutive points on the x axis, which is 1/2 in this example. Now consider Table 6.4. The approximate area of the bounded region is $140/16 = 8.75$. The value 8.75 is the sum of the areas of the

FIGURE 6.26

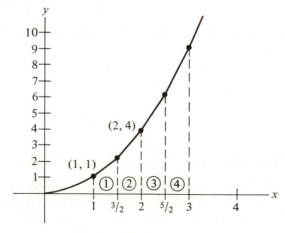

TABLE 6.4

Trapezoid	Altitude	Base 1	Base 2	Area
1	1/2	$f(1) = 1$	$f(3/2) = 9/4$	13/16
2	1/2	$f(3/2) = 9/4$	$f(2) = 4$	25/16
3	1/2	$f(2) = 4$	$f(5/2) = 25/4$	41/16
4	1/2	$f(5/2) = 25/4$	$f(3) = 9$	61/16

four trapezoids. We can compare this approximated area of 8.75 with the area computed by the integral.

$$A = \int_1^3 x^2 \, dx = \left. \frac{x^3}{3} \right|_1^3$$

$$F(3) = 9$$

$$F(1) = \frac{1}{3}$$

$$F(3) - F(1) = 9 - \frac{1}{3} = 8.666 \ldots$$

By subtracting the two areas, an error of .08333 . . . is found in the approximated method. Could this error be reduced? Yes, the only required action would be to divide the interval $1 \leq x \leq 3$ into more equal divisions and thus obtain more trapezoids. This becomes increasingly difficult to compute by hand, but a computer does not mind doing these messy manipulations.

Figure 6.27 shows the interval divided into eight equal divisions instead of

FIGURE 6.27

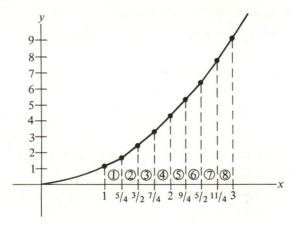

four. The area approximated by the eight trapezoids is 8.6875. Comparing this approximated area with the area found by integration $(8.6875 - 8.666\ldots)$ yields a difference of .02083. . . . This difference shows that a better approximation was found by taking eight equal intervals. This approximation can be made more accurate by taking more equal intervals.

Summarizing the material into a formula for approximating the area over $a \leq x \leq b$, we note the following entries in Table 6.4. First, the altitude is the same for every trapezoid. This altitude can be computed by taking $(b - a)/n$, where n is the number of equal intervals. Second, note that each base is used twice except for the base whose length is $f(a)$ and the base whose length is $f(b)$. Thus we can summarize for a nonnegative function by the following statement:

$$\int_a^b f(x)\, dx \doteq \tfrac{1}{2} \cdot \frac{b - a}{n} \{ f(a) + f(b) + 2$$

$$\times [f(x_1) + f(x_2) + f(x_3) + f(x_4) + \cdots + f(x_{n-1})]\}$$

where $x_1, x_2, x_3, x_4, \ldots, x_{n-1}$ are the points on the x axis between a and b. These points are determined by the number of intervals n.

Example 28 Approximate the area bounded by $f(x) = 1/x$, the x axis and the lines $x = 1$, $x = 5$. Let $n = 8$.

The altitude is $(5 - 1)/8 = \tfrac{1}{2}$. The values for the points on the x axis are $a = 1$, $x_1 = \tfrac{3}{2}$, $x_2 = 2$, $x_3 = \tfrac{5}{2}$, $x_4 = 3$, $x_5 = \tfrac{7}{2}$, $x_6 = 4$, $x_7 = \tfrac{9}{2}$, and $b = 5$. After computing $f(x)$ for each of the eight values, the area can be approximated:

$$A \doteq \tfrac{1}{4}[1 + \tfrac{1}{5} + 2(\tfrac{2}{3} + \tfrac{1}{2} + \tfrac{2}{5} + \tfrac{1}{3} + \tfrac{2}{7} + \tfrac{1}{4} + \tfrac{2}{9})] = 1.6290$$

▶ **Exercise 6.6**

Approximate the area in each of the following problems by means of the trapezoidal rule. In each problem use $n = 4$.

1 Find the approximate area of the region bounded by $f(x) = x^3$, the x axis, the vertical lines $x = 1$, $x = 3$.

2 Find the approximate area of the region bounded by $f(x) = \sqrt{x}$, the x axis, and the vertical lines $x = 0$, $x = 1$.

3 Find the approximate area of the region bounded by $f(x) = 1/(x - 1)$, the x axis, and the vertical lines $x = 2$, $x = 4$.

4 Find the approximate area of the region bounded by $f(x) = x^2 - 2x$, the x axis, and the vertical lines $x = 3$, $x = 4$.

5 Find the approximate area of the region bounded by $f(x) = x + 1/x$, the x axis, and the vertical lines $x = 2$, $x = 4$.

6 Find the approximate area of the region bounded by $f(x) = e^{-x^2}$, the x axis, and the vertical lines $x = -1$, $x = 1$.

7 Find the area in Problem 1 by using the definite integral. Compute the error in the approximation.

8 Find the area in Problem 2 by using the definite integral. Compute the error in the approximation.

9 Find the area in Problem 4 by using the definite integral. Compute the error in the approximation.

---

self-test · chapter six

Find the value of the definite integral in Problems 1 through 3.

1 $\displaystyle\int_1^2 (x^2 - x + 3)\, dx$ **2** $\displaystyle\int_0^2 e^{2x}\, dx$ **3** $\displaystyle\int_1^4 \frac{1}{\sqrt{x}}\, dx$

4 Find the area of the region bounded by $f(x) = -x^2 + 6$, the x axis, and the vertical lines $x = 1$, $x = 2$.

5 Find the area of the region between $f(x) = x + 7$ and $g(x) = x^2 + 1$.

6 Find the value of the following improper integral, if possible:

$$\int_1^\infty \frac{1}{x^4}\, dx$$

7 Let the demand function be given by $f(x) = -x^2 + 2x + 5$. The fixed price is given by $g(x) = 4$. Find the consumer surplus.

8 Approximate the area bounded by $f(x) = 1/(x + 1)$, the x axis, and the two vertical lines $x = 2$, $x = 3$. Use $n = 4$.

functions of several variables

7.1 Higher Derivatives

A higher derivative is the derivative of a function that is itself a derivative. For example, if $y = f(x)$, then $y' = f'(x)$ is the *first derivative* of $y = f(x)$. The derivative of $y' = f'(x)$ is called the *second derivative* of $y = f(x)$ and is denoted $y'' = f''(x)$. The derivative of the second derivative is called the *third derivative* of $y = f(x)$ and is denoted $y''' = f'''(x)$, and so on.

Example 1   Let $y = 3x^3 - 6x - 12$. Find the first, second, and third derivatives of the function.

$$y' = 9x^2 - 6$$
$$y'' = 18x$$
$$y''' = 18$$

Example 2   Let $f(x) = (x + 1)/x$. Find the first, second, and third derivatives of the function.

$$f'(x) = \frac{-1}{x^2}$$

$$f''(x) = \frac{2}{x^3}$$

$$f'''(x) = \frac{-6}{x^4}$$

Applied Example 3   Suppose the distance an object moves in a straight line in time t is given by $s(t) = t^3 + t^2 + t$. Find the first and second derivatives.

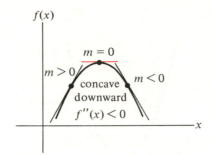

FIGURE 7.1

$$s'(t) = 3t^2 + 2t + 1 \quad \text{(velocity function)}$$
$$s''(t) = 6t + 2 \qquad \text{(acceleration function)}$$

The usefulness of the first derivative has already been demonstrated. Of the higher derivatives, only the second is of significant value in this course.

Let us examine the geometric interpretation of the second derivative. The first derivative determines the slope of the tangent line to the graph of a given function at a point. The second derivative is the rate of change of the first derivative. Therefore if the second derivative is negative, the tangent lines have decreasing slopes. For the values of x where the second derivative is negative, the graph of the function is said to be *concave downward*. (See Figure 7.1.)

If the second derivative is positive, the tangent lines have increasing slopes. For the values of x where the second derivative is positive, the graph of the function is said to be *concave upward*. (See Figure 7.2.)

Example 4 What is the concavity of $f(x) = x^2$?

Since $f'(x) = 2x$, $f''(x) = 2$, which is positive for all values of x. Therefore the graph of the function is concave upward. (See Figure 7.3.)

Example 5 What is the concavity of $y = 2 - x - x^2$?

Since $y' = -1 - 2x$, $y'' = -2$, which is negative for all values of x. Therefore the graph of the function is concave downward. (See Figure 7.4.)

A change in concavity sometimes occurs when the second derivative is zero.

FIGURE 7.2

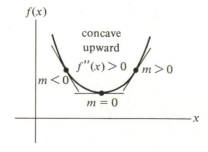

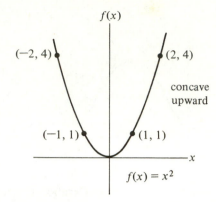

FIGURE 7.3

Example 6 Let $y = x^3 - 4x^2 + 7x - 5$.

1. Find the value of x where the second derivative is zero.

2. Determine the concavity of the function to the left and right of the value of x found in part 1.

$$y' = 3x^2 - 8x + 7$$

$$y'' = 6x - 8$$

Set $6x - 8 = 0$; then $x = \frac{4}{3}$.

To determine the concavity to the right of $x = \frac{4}{3}$, let $x = 2$. Then $y'' = 4$, which is greater than zero. Therefore the function is concave upward to the right of $x = \frac{4}{3}$ (that is, $y'' > 0$ for $x > \frac{4}{3}$).

To determine the concavity to the left of $x = \frac{4}{3}$, let $x = 0$. Then $y'' = -8$. Therefore the function is concave downward to the left of $x = \frac{4}{3}$ (that is, $y'' < 0$ for $x < \frac{4}{3}$). (See Figure 7.5.)

FIGURE 7.4

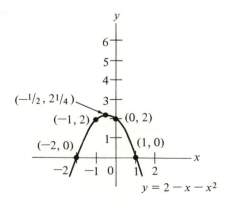

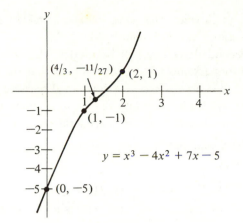

FIGURE 7.5

Example 7   Let $y = x^3/3 - 2x^2 + 3x + 2$.

1. Find the values of x where the slope of the tangent line to the curve is zero.

2. Find the value of x where the curve changes concavity.

$$y' = x^2 - 4x + 3$$

Set $x^2 - 4x + 3 = 0$ and solve. Then $x = 3$ and $x = 1$ are the values of x where the slope of the tangent lines to the curve is zero.

$$y'' = 2x - 4$$

Set $2x - 4 = 0$ and solve. Then $x = 2$ is the value of x where the concavity changes. (See Figure 7.6.)

FIGURE 7.6

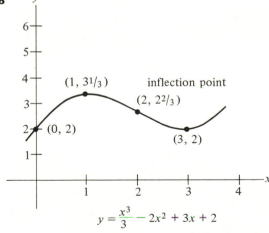

The point on the graph where the curve changes concavity is called an *inflection point*. The inflection point in Example 7 is $(2, 2\frac{2}{3})$.

Sometimes the second derivative can be used to determine maximum and minimum points. From Example 7 it can be noted that the point $(1, 3\frac{1}{3})$ is a local maximum and the curve is concave downward because $f''(1)$ is negative. The point $(3, 2)$ is a local minimum and the curve is concave upward because $f''(3)$ is positive. These observations can be generalized as follows:

Finding maximums and minimums (*second derivative test*)

1 Find $f'(x)$ and $f''(x)$.

2 If $f'(p) = 0$ and $f''(p) > 0$, the point $(p, f(p))$ is a minimum.

3 If $f'(p) = 0$ and $f''(p) < 0$, the point $(p, f(p))$ is a maximum.

4 If $f'(p) = 0$ and $f''(p) = 0$, the second derivative test fails and one of the original tests for maximums and minimums should be employed.

Example 8 Find the maximum and minimum points of $y = x^3/3 - 9x + 10$.

$$y' = x^2 - 9 = 0$$

Therefore $x = -3$ and $x = 3$. Now $y'' = 2x$; thus at $x = -3$, $y'' = -6 < 0$. Hence $(-3, 28)$ is a maximum. At $x = 3$, $y'' = 6 > 0$, and the point $(3, -8)$ is a minimum. (See Figure 7.7.)

Applied Example 9 A company has determined that its profit in dollars is related to the amount of money in hundreds of dollars spent on advertising, as given by $P(x) = 1000 + 30,000x - x^3$. Find the value of x that yields the maximum profit. Find the maximum profit.

$$P'(x) = 30,000 - 3x^2 = 0$$

so that $x = 100$. (The value $x = -100$ is not meaningful.)

$$P''(x) = -6x.$$

FIGURE 7.7

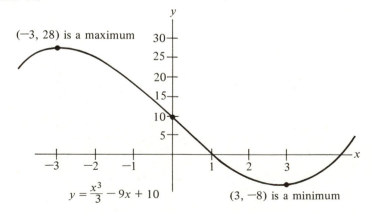

$(-3, 28)$ is a maximum

$y = \dfrac{x^3}{3} - 9x + 10$

$(3, -8)$ is a minimum

Since $P''(100)$ is negative, the point $(100, 2{,}001{,}000)$ is a maximum point. The maximum profit is $2,001,000.

Nonexample 10   Let $y = x^4$. Find the maximum and minimum points.

$$y' = 4x^3 = 0$$

Therefore $x = 0$. But $y'' = 12x^2 = 0$ at $x = 0$. Therefore the second derivative test fails, but the point $(0, 0)$ is a minimum, because $y'(-1) = -4 < 0$ and $y'(1) = 4 > 0$.

▶ **Exercise 7.1**

Find the first and second derivatives of the functions in Problems 1 through 8.

1 $y = 6x^5 + 4x^3 + 1$

2 $f(x) = x^3 + 4x^2 - 6x$

3 $c(n) = 2n^3 - 8n^2 + n$

4 $y = \dfrac{1}{x}$

5 $y = \sqrt{x+1}$

6 $R(x) = 4x^3 + \dfrac{1}{x}$

7 $y = e^x$

8 $y = e^{-2x}$

9 If $s(t) = t^2 + 2/t$ is a distance function, find the velocity and acceleration functions.

Find the maximum and/or minimum points of the functions in Problems 10 through 16. Use the second derivative test when possible.

10 $y = 3x^2 + 18x + 1$

11 $y = 10 - 6x - x^2$

12 $y = x^2 + 6x + 5$

13 $q = 2 - n - 3n^2$

14 $y = x^3 - 9x^2$

15 $y = x^6$

16 $y = -e^{x^2}$

17 A company knows that its average cost depends on the number of units x produced, as given by $A(x) = 1000 + 2x^2 - 80x$. Find the minimum average cost.

18 Suppose it costs $1000 to prepare for a production run of a certain item. Once production is under way, it costs $5 to produce each item. After the items are produced, it costs an average of $2 per year to store each item until it is needed. The company needs 1000 items per year. Find the number of units that should be produced during each run to minimize the total cost.

Assume that all the items produced during each production run are stored and withdrawn at a constant rate and that each run is depleted before the next production run is initiated. We can represent the average number of units in storage as $x/2$, where x is the number of units produced by each

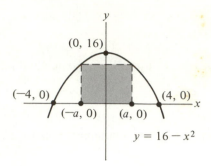

FIGURE 7.8

production run, and the cost of storage as $2x/2 = x$. Thus the total cost equation for this problem is $C = 1000(1000/x) + 5(1000) + x$.

19 Find two positive numbers x and y such that their sum is 64 and their product is a maximum.

20 Find the maximum and/or minimum point(s) of $y = e^{x^2-2}$.

21 Find the dimensions of the largest rectangle that can be drawn with one side on the x axis and two of its vertices on the curve $y = 16 - x^2$. (See Figure 7.8.)

22 Suppose it costs $2000 to prepare for a production run. Once production is under way, it costs $10 to produce each unit. If the average cost of storing each item is $5 per year and if the company needs 5000 units per year, how many units should be produced during each production run to minimize cost?

23 If the cost per item is removed from Problem 22, the cost can be written as $C(x) = (x/2) \cdot c + (d/x) \cdot S$, where c is the carrying cost, d is the demand, and S is the amount required to set up each production run and x is the number of units produced in each production run. If c, d, and S are constant, prove that the minimum cost occurs at $x = \sqrt{(2dS)/c}$.

24 Find the number of units that will minimize cost if $c = .25$, $d = 10,000$, and $S = 200$.

7.2 Functions of Several Variables

In previous sections each function has contained only one independent variable. For example, total cost has depended solely on the number of units produced. We have used simplified functions of one independent variable initially in order to illustrate mathematical concepts and develop mathematical acuity.

However, in a more realistic situation the total cost depends on more than production. It may depend on units x produced, transportation t, advertising a, salaries s, maintenance m, and interest on capital outlay i. Therefore total cost C is a function of x, t, a, s, m, and i, denoted by $C = f(x, t, a, s, m, i)$. In this case, cost is a function of six variables. To determine the effect of each independent variable on the dependent variable, we hold all variables constant except the one being analyzed.

Example 11 Let $z = f(x, y) = 2x + 3y$. To understand that the value of z depends on both the value of x and the value of y, consider Tables 7.1 and 7.2.

TABLE 7.1

(x, y)	z
$(1, 1)$	5
$(1, 2)$	8
$(1, -3)$	-7
$(1, 0)$	2

TABLE 7.2

(x, y)	z
$(1, 1)$	5
$(2, 1)$	7
$(3, 1)$	9
$(-4, 1)$	-5
$(0, 1)$	3

If we hold x constant, the value of z changes as the value of y changes. If we hold y constant, the value of z changes as the value of x changes.

If the value of an independent variable changes, the corresponding value of the dependent variable changes, as shown in Example 11. By using the concept of the derivative we can determine more than changes in functional values. The function can be analyzed as to the *rate* of *change* of the dependent variable with respect to *one* of the independent variables. This is done by holding all independent variables constant except the one whose effect is being analyzed and differentiating, using the rules from Chapters 3 and 4. When all but one of the independent variables are held constant, the derivative is called a *partial derivative*, which we abbreviate as *partial*.

If $z = f(x, y)$, there are two first partials. One partial is determined by holding x constant and letting y be the variable. This is called the *first partial of z with respect to y* and is denoted z_y or $f_y(x, y)$. Another partial can be found by holding y constant and letting x be the variable. This is called the *first partial of z with respect to x* and is denoted z_x or $f_x(x, y)$.

Example 12 Let $z = 2x^2 + 3xy - 6$. Find both first partials.

$$z_x = 4x + 3y$$
$$z_y = 3x$$

Example 13 Let $z = 2x^2y^3 + 7xy^3$. Find both first partials.

$$z_x = 4xy^3 + 7y^3$$
$$z_y = 6x^2y^2 + 21xy^2$$

Example 14 Let $f(x, w, n) = 2x^2 - 3nx - w^2x$. Find all first partials.

$$f_x(x, w, n) = 4x - 3n - w^2$$
$$f_w(x, w, n) = -2xw$$
$$f_n(x, w, n) = -3x$$

As with functions of one independent variable, the first partials can be differentiated again, yielding second partial derivatives. Several possibilities exist for finding second partials. After getting the first partial with respect to x, denoted $f_x(x, y)$, the second partial can be determined with respect to x again or with respect to y, denoted $f_{xx}(x, y)$ and $f_{xy}(x, y)$, respectively. If the first partial is determined with respect to y, denoted $f_y(x, y)$, the second partial can be determined with respect to y again or with respect to x, denoted $f_{yy}(x, y)$ and $f_{yx}(x, y)$, respectively.

Example 15 Let $z = f(x, y) = 2xy^2 - 6x^3y^5$. Find all first and second partials.

$$z_x = 2y^2 - 18x^2y^5 \qquad z_{xx} = -36xy^5$$
$$z_{xy} = 4y - 90x^2y^4$$
$$z_y = 4xy - 30x^3y^4 \qquad z_{yy} = 4x - 120x^3y^3$$
$$z_{yx} = 4y - 90x^2y^4$$

Note that $z_{xy} = z_{yx}$. This is true when the derivatives are continuous functions.

Applied Example 16 A firm's cost depends on the number of units produced x and the number of employees y, as given by $C = 6x + 3y + 10xy$. Find the rate of change of cost with respect to production.

$$C_x = 6 + 10y$$

Find the rate of change of cost with respect to the number of employees.

$$C_y = 3 + 10x$$

Applied Example 17 Robert Gunning, in his work "How To Take the Fog Out of Writing"* devised the fog index to help eschew obfuscation in written prose. Prose has a large fog index if it contains lengthy sentences and multitudinous polysyllabic words, a small fog index if it contains succinct sentences and a low frequency of occurrence of polysyllabic words. The fog index approximates the number of years of matriculation requisite for comprehension of the

---

* Robert Gunning, *The Technique of Clear Writing*, rev. ed. (McGraw-Hill, Inc., New York, 1968), p. 9–10. Copyright © by Robert Gunning. Reprinted with permission.

composition. The fog index can be computed from a 100-word excerpt with exiguous manipulation by implementation of the formula

$$F = .4\left(\frac{100}{x} + \frac{y}{100}\right)$$

where x is the number of sentences and y is the number of polysyllabic words.
    Determine the rate of change of F with respect to x.

$$F_x = .4\left(\frac{-100}{x^2}\right) = \frac{-40}{x^2}$$

The rate of change of the fog index varies inversely with the square of the number of sentences. That is, the more sentences, the clearer the composition (within reasonable limits).
    Determine the rate of change of F with respect to y.

$$F_y = .4\left(\frac{1}{100}\right) = .004$$

The fog index varies constantly with the number of polysyllabic words.

▶ **Exercise 7.2**

Find all first and second partials.

1 $C = 2x^2 - 3n^3 + xn$ 2 $z = 2x^2y^5 - x^3y + 6$

3 $z = 2x^2y + x^{1/2}$ 4 $z = 2xy + 3x^2y + xy^3 - 10$

5 $z = \dfrac{x}{y^2} - xy$ 6 $R = xy - \dfrac{1}{x} + \dfrac{1}{y}$

7 $z = e^{2x} + xy$ 8 $z = e^p + 2p^2q^3$

9 The atmospheric pressure varies with both height h and temperature t, as given by the equation $P = e^{-kh/t}$. Find the rate of change of pressure with respect to h and with respect to t (k is constant).

10 Suppose the production of a certain product requires an input of labor l and an input of material m. Then the production z is a function of l and m, denoted $z = f(l, m)$. If $z = f(l, m) = 2l^3 - lm^2 + 3lm$, find the rate of change of production with respect to each input.

11 If the demand D for a certain commodity depends on its market price p and the number of units n available, as represented by $D = n^2 - 3np + p^2$, find the expression for the rate of change of demand with respect to price and with respect to the number of units available.

12 Find the fog index of the beginning of Chapter 1. Find the fog index of Applied Example 17.

13 The force of the gravitational attraction between two objects is $f = gm_1m_2/d^2$,

where m_1 and m_2 represent the masses of the two objects, d is the distance between the objects, and g is the gravitational constant. Find the rate of change of f with respect to each of the variables.

14 In electrical theory we have the equation $I = V/R$, where R is the resistance to the flow of electricity, V is the electromotive force, and I is the current. Find the rate of change of I with respect to each of the other variables.

7.3 The Maximum and Minimum Values of a Function of Two Independent Variables

Finding the maximum and/or minimum values of a function of two variables requires the use of all the first and second partials. As with functions of one independent variable, the maximum and minimum values may be located when the first derivative (first partials) is equal to zero. Also, as with functions of one independent variable, the first partials being equal to zero is not sufficient to guarantee that the function has a maximum or minimum value. To find the maximum or minimum value we must first determine the point (a, b) at which the maximum or minimum value may occur. The procedure for finding maximum and minimum values is as follows.

Let $z = f(x, y)$ be a function of two independent variables whose first and second partials exist.

1 Find $f_x(x, y)$ and $f_y(x, y)$. Set $f_x(x, y) = f_y(x, y) = 0$. Then solve for their simultaneous solution(s), if there are any. If there is a solution (a, b), go to step 2.

2 Find $f_{xx}(x, y)$, $f_{yy}(x, y)$, and $f_{xy}(x, y)$. Compute
$$T = f_{xx}(a, b) \cdot f_{yy}(a, b) - [f_{xy}(a, b)]^2$$

If $T < 0$, neither a maximum nor a minimum occurs at (a, b).
If $T > 0$, either a maximum or minimum occurs at (a, b). (Go to step 3.)
If $T = 0$, the test fails.

3 If $f_{xx}(a, b) < 0$, then (a, b) yields a local maximum. If $f_{xx}(a, b) > 0$, then (a, b) yields a local minimum.

Example 18   Let $z = x^2 + 2y^2 - 8y$. Find all maximum and minimum values, if there are any.

1. $z_x = 2x$. Set $2x = 0$; then $x = 0$.

   $z_y = 4y - 8$. Set $4y - 8 = 0$; then $y = 2$.

Therefore $(a, b) = (0, 2)$.

2. $z_{xx} = 2$, $z_{yy} = 4$, and $z_{xy} = 0$, $T = (2)(4) - 0^2 > 0$, so $(0, 2)$ is a maximum or a minimum point.

3. $z_{xx} = 2 > 0$. Therefore $(0, 2)$ yields a minimum. The minimum value is $z = (0)^2 + 2(2)^2 - 8(2) = -8$.

Example 19   Let $z = xy - 2/x - 4/y$. Find the maximum and minimum values if there are any.

1.
$$z_x = y + \frac{2}{x^2} = 0$$

$$z_y = x + \frac{4}{y^2} = 0$$

To solve z_x and z_y simultaneously, solve $z_x = 0$ for y: $y = -2/x^2$. Now substitute for y in $x + 4/y^2 = 0$. Hence

$$x + \frac{4}{(-2/x^2)^2} = 0$$

$$x + \frac{4}{(4/x^4)} = 0$$

$$x + x^4 = 0$$

Factoring, $x(x^3 + 1) = 0$. Setting each factor equal to zero and solving for the real roots yields $x = 0$ and $x = -1$. At $x = 0$, y is undefined. At $x = -1$, $y = -2$. Hence $(a, b) = (-1, -2)$.

2.
$$z_{xx} = \frac{-4}{x^3}$$

$$z_{yy} = \frac{-8}{y^3}$$

$$z_{xy} = 1$$

$$T = \frac{-4}{(-1)^3} \cdot \frac{-8}{(-2)^3} - (1)^2 = 3 > 0$$

Therefore $(-1, -2)$ yields a maximum or a minimum value.

3. $z_{xx} = -4/(-1)^3 = 4 > 0$, so $(-1, -2)$ yields a minimum. The minimum value is $z = 6$.

Nonexample 20   Let $z = 9xy$. Find all maximum and minimum values, if there are any.

1.
$$z_x = 9y = 0; \text{ hence } y = 0$$
$$z_y = 9x = 0; \text{ hence } x = 0$$

2. $$z_{xx} = 0; z_{yy} = 0; \text{ and } z_{xy} = 9; T = 0 - 9^2 = -81$$

Therefore the function has no maximum or minimum value.

Applied Example 21 A company has determined that its profit P depends on the amount m spent on advertising and the cost of labor l, as given by

$$P = f(m, l) = ml - m^2 - l^2 + 80m + 20l + 1000$$

Find the combination of advertising and labor (m, l) that yields the maximum profit.

1. $$P_m = l - 2m + 80$$
$$P_l = m - 2l + 20$$

Solving simultaneously, $m = 60$ and $l = 40$. So $(a, b) = (60, 40)$.

2. $P_{mm} = -2$, $P_{ll} = -2$, and $P_{ml} = 1$. Hence $T = 3 > 0$. Therefore $(60, 40)$ yields either a maximum or a minimum value.

3. Since $P_{mm} = -2 < 0$, $(60, 40)$ yields a maximum value. The maximum value is $P = 3800$.

▶ **Exercise 7.3**

Find all points at which maximum and/or minimum values occur in the functions of Problems 1 through 5, if there are any. Find the maximum or the minimum values.

1 $z = f(x, y) = x^2 + 4x + y^2 - 10y$

2 $z = f(x, y) = x^2 - x + y^2 - 2y + 1$

3 $z = f(x, y) = 2xy$

4 $z = f(x, y) = 2x^2 - 16x - y^2 + 3$

5 $z = f(x, y) = x^3 + y^3$

6 A company knows that its cost $C(x, y)$ depends on the number of retail outlets x and the number of delivery trucks y, as given by $C(x, y) = x^3 - 10xy + y^3 + 10,000$. Find the number of outlets and trucks needed to minimize cost.

7 A company produces an item that requires m man-hours and n machine-hours to assemble. If the total assembly cost z is given by $z = 2n^3 - 12mn + m^2 + 2728$, find the combination of man-hours and machine-hours to produce the item at minimum cost.

7.4 Differentiation of Implicit Equations Using Partials

The equations introduced so far have been explicit equations. Each equation has been written as $y = f(x)$ or as $z = f(x, y)$, where the dependent variable is explicitly a function of the independent variable(s).

Example 22 Each of the following is an explicit equation.

1 $y = x^2 - 2x - 3$ 2 $R(n) = 2n + n^3$

3 $C(x) = x^3 + x - 6$ 4 $z = x^3 + y^3$

5 $f(x, y) = 2x^2 - 16x - y + 3$

An *implicit equation* is one that is *not* explicitly solved for the dependent variable in terms of the independent variable(s).

Example 23 Each of the following is an implicit equation.

1 $2x^2 + 3xy - y^2 = 6$ 2 $xy = 1$

3 $\dfrac{x}{y^2} - 2xy = 0$ 4 $y^2 = 2x + 1$

5 $z - x^3 - y^3 = 0$ 6 $x^2 + y^2 + z^2 = 1$

To find the derivative dy/dx, the implicit equation may be solved for the dependent variable and the derivative determined as usual.

Example 24 Let $x^2 - 2x - y = 5$. Find dy/dx.

Solving the equation for y, we have $y = x^2 - 2x - 5$. Therefore $dy/dx = 2x - 2$.

However, it may be difficult (or impossible) to solve for y. In this case, the derivative may be obtained by the use of partials, as shown in Rule 9.

RULE 9

If the equation $F(x, y) = 0$ defines an implicit function $y = f(x)$, then

$$\frac{dy}{dx} = \frac{-F_x(x, y)}{F_y(x, y)}$$

provided the partials are continuous and $F_y(x, y) \neq 0$.

Example 25 Let $2x^2 + 3xy - y^2 = 6$. Find dy/dx.

Let $F(x, y) = 2x^2 + 3xy - y^2 - 6 = 0$. Then the partials are $F_x(x, y) = 4x + 3y$ and $F_y(x, y) = 3x - 2y$. Therefore

$$\frac{dy}{dx} = \frac{-(4x + 3y)}{3x - 2y}$$

Example 26 Let $(1/x) + (2xy) + (1/y) = 5$. Find dy/dx.

Let $F(x, y) = 1/x + 2xy + 1/y - 5 = 0$. Then the partials are $F_x(x, y) = -1/x^2 + 2y$ and $F_y(x, y) = 2x - 1/y^2$. Therefore

$$\frac{dy}{dx} = \frac{-(-1/x^2 + 2y)}{2x - 1/y^2}$$

$$= \frac{1/x^2 - 2y}{2x - 1/y^2}$$

$$= \frac{1 - 2x^2y}{x^2} \cdot \frac{y^2}{2xy^2 - 1}$$

$$= \frac{y^2 - 2x^2y^3}{2x^3y^2 - x^2}$$

▶ **Exercise 7.4**

Find dy/dx in Problems 1 through 4.

1 $2x^2 + 3y^2 = 6$

2 $x^2 + 6xy + y^2 = 0$

3 $2x^3 = 6x + 2y - xy$

4 $\dfrac{1}{x} + xy - y^3 = 2x + 1$

5 If $x^2 + xy = 2$, find the slope of the tangent line to the curve at $(1, 1)$. Write the equation of the tangent line to the curve at $(1, 1)$.

6 The selling price x in dollars of a specific item is related to the demand y in units for that item by $100x + xy + 2y - 4600 = 0$.
   a What number of units must be supplied to satisfy the market demand if the price is $10?
   b What is the rate of change of demand with respect to the selling price?
   c At a selling price of $10, is the demand for the item increasing?

7.5 Lagrangian Multipliers

During the eighteenth century the French mathematician Joseph Lagrange developed a method for finding the maximum or minimum value of a constrained function of two or more independent variables. A *constraint* is a restriction placed on the variables in addition to those implied in the objective function.

To present his method we will use the following terminology:

1 The function to be maximized or minimized will be called the *objective function* and will be denoted $z = f(x, y)$.

2 The constraint will be denoted $g(x, y) = 0$.

3 The Lagrange multiplier is an independent variable denoted k, from which this method gets its name.

4 The expressions above will be combined into an equation called the Lagrange equation, denoted $L = h(x, y, k)$.

The first step in forming the Lagrange equation is to add the objective function to the constraint multiplied by k, yielding

$$L = h(x, y, k) = f(x, y) + k[g(x, y)]$$

Then all first partials of L are found and set equal to zero, and the resulting system of equations is solved simultaneously for the values of x, y, and k. The point (x, y) is called a critical point. This point (x, y) is then tested to see if a maximum or a minimum value (or neither) occurs at (x, y).

For the technique to work, the objective function must have a maximum or a minimum value at (a, b), and the functions $f(x, y)$ and $g(x, y)$ must be continuous for all values of x and y near (a, b). The technique is illustrated in Examples 27 and 28.

Applied Example 27   A rectangular area is to be fenced adjacent to a building. The total amount of fence to be used is 30 ft. Find the dimensions of the rectangle that yield the maximum area.

The objective function is the area of the rectangle. If we denote the area z and the dimensions x and y, where x is parallel to the building, the objective function is $z = f(x, y) = xy$.

The constraint is the 30 ft of fence available for the perimeter. The constraint is $g(x, y) = x + 2y - 30 = 0$.

Next, form the Lagrange equation

$$L = h(x, y, k) = f(x, y) + k[g(x, y)]$$
$$= xy + k(x + 2y - 30)$$

The first partials of L are

1. $L_x = y + k$
2. $L_y = x + 2k$
3. $L_k = x + 2y - 30$

Solving simultaneously,

1. $y + \ k = 0$
2. $x + 2k = 0$

Multiplying the first equation by -2 and eliminating k yields $x - 2y = 0$.

Solving $x - 2y = 0$ simultaneously with the constraint $x + 2y = 30$, we find that $x = 15$. When $x = 15$, then $15 - 2y = 0$, so $y = 7.5$. Also, when $x = 15$

and $y = 7.5$, then $k = -7.5$. Therefore the point $(15, 7.5)$ is a critical point. $L = h(15, 7.5, -7.5) = 112.5$.

To determine if 112.5 is a maximum or a minimum value of the function $z = f(x, y)$ at $(15, 7.5)$, we must choose two other "triplets" of numbers, changing the values of x and y while holding k fixed at -7.5. The test values of x and y must be reasonably close to 15 and 7.5 and *must satisfy the constraint*. Compute the values of L. If the two values of L are larger than 112.5, then 112.5 is a minimum. If the two values of L are smaller than 112.5, then 112.5 is a maximum.

Choose the two triples $(14, 8, -7.5)$ and $(16, 7, -7.5)$.

$$L = h(14, 8, -7.5) = 112$$
$$L = h(16, 7, -7.5) = 112$$

Since both values of L are smaller than 112.5, 112.5 is a maximum value of $z = f(x, y)$. That is, $z = f(15, 7.5) = 112.5$ is the maximum area of the rectangular area to be fenced.

In Example 28 we use the method of Lagrange multipliers to solve for the constrained maximum or minimum value of a function that has three independent variables with one constraint. (The method can be extended to solve problems that have several independent variables and several constraints. However these cases will not be considered here.)

Example 28 Find the minimum value of $w = x^2 + y^2 + z^2$, subject to the constraint $x + y + z = 3$.

$$L = x^2 + y^2 + z^2 + k(x + y + z - 3)$$
$$L_x = 2x + k$$
$$L_y = 2y + k$$
$$L_z = 2z + k$$
$$L_k = x + y + z - 3$$

Solving the system $L_x = L_y = L_z = L_k = 0$, we observe that $2x + k = 2z + k$, so $x = z$. Also, since $2x + k = 2y + k$, $x = y$. Therefore $x = y = z$.

Substituting into $L_k = 0$, we have

$$x + x + x - 3 = 0$$
$$3x - 3 = 0$$
$$3x = 3$$
$$x = 1$$

Therefore $y = 1$, $z = 1$, and $k = -2$. The critical point is $(1, 1, 1)$.

To test this point we choose some nearby points. Choose $(2, 1, 0, -2)$ and $(0, 3, 0, -2)$. *Be careful.* The constraint requires that $x + y + z = 3$. Points such as $(0, 0, 2, -2)$ will not suffice.

$$L = 1^2 + 1^2 + 1^2 - 2(1 + 1 + 1 - 3) = 3$$

at the critical point.

$$L = 2^2 + 1^2 + 0^2 - 2(2 + 1 + 0 - 3) = 5$$

$$L = 0^2 + 3^2 + 0^2 - 2(0 + 3 + 0 - 3) = 9$$

at the checkpoints. Since L has larger values at the checkpoints than it does at the critical point, the critical point is $(1, 1, 1)$, and 3 is the minimum value of $w = x^2 + y^2 + z^2$.

▶ **Exercise 7.5**

Find the point at which the constrained maximum or minimum occurs in Problems 1 through 6. Find the maximum or minimum value.

1 $z = x^2 + 3xy + y^2$, constraint $x + y = 30$

2 $z = x^2 - xy + y^2$, constraint $x + y = 20$

3 $z = x^2 + 2xy + y^2$, constraint $x - 3y = 20$

4 $q = x^2 + y^2 + z^2$, constraint $x + y - z = 3$

5 $q = 2x^2 + y^2 + z^2$, constraint $x + 3y + 6z = 9$

6 $q = x^2 + y^2 - z^2$, constraint $x + 2y + 2z = 8$

7 $z = x^2 + 2xy + y^2$, constraint $x + 2y = 20$.

8 A company produces two items A and B. The cost of producing these items for any 6-hr shift is given by $C = 2x^2 + 20y$, where x is the number of machines producing item A and y is the number of machines producing item B. If there are exactly 20 machines in operation during each shift, find the number of machines producing item A and the number of machines producing item B that will minimize cost. What is the minimum cost?

9 A company has three plants producing a certain item. The company would like to fill an order for 700 units in the most economical way. If the production cost for each plant is

$$C_1(x) = 300 + x^2; \quad C_2(y) = 200 + 4y^2; \quad C_3(z) = 200 + 2z^2$$

find the optimum way production should be allocated to minimize production cost. (*Hint:* Add the individual cost functions to get the total cost.)

self-test · chapter seven

Find the indicated derivatives in Problems 1 through 5.

1 $y = x^2 + 6x - 7$

$y' =$

$y'' =$

2 $y = x + 1/x$

$y' =$

$y'' =$

3 $y = e^{x^2} + x^3 + 5$

$y' =$

$y'' =$

4 $x + xy^3 + y^2 = 10$

$\cdot dy/dx =$

5 $e^x y + x e^y = 1$

$dy/dx =$

6 Find the indicated partials:

$$z = x^2 + y^2 + 3xy^3$$

$$z_x =$$

$$z_{xx} =$$

$$z_{xy} =$$

$$z_y =$$

$$z_{yy} =$$

Find the maximum and/or minimum points in Problems 7 and 8.

7 $z = 2x^2 + 6y^2 + 12y$

8 $z = 10x + 2y + xy - x^2 - .5y^2$

9 Find the constrained maximum or minimum: $z = x^2 - 6xy - y^2$ constrained to $x + v - 6 = 0$.

logarithms*

8.1 Logarithmic Functions

Recall that graphing of exponential equations was introduced in Chapter 1. To graph a logarithmic equation, such as $y = \log_2 x$, we may rewrite the logarithmic equation in exponential notation and proceed as before.

Example 1   Graph $y = \log_2 x$. (See Figure 8.1.)

Rewriting, $x = 2^y$. The x intercept is 1. There is no y intercept, since no value of y makes 2^y equal to zero. Also, the equation $y = \log_2 x$ is a function, because for each positive value of x, there is a single (unique) value of y.

Although x is the independent variable in this example and y is the dependent

FIGURE 8.1

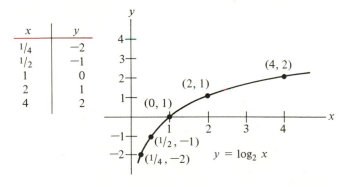

* This chapter assumes a knowledge of the basic properties of a logarithm, which are discussed in Appendix 2.

x	y
.135	−2
.368	−1
1	0
2.718	1
7.389	2

(7.389, 2)

(2.718, 1)

(0, 1)

(.368, −1)

(.135, −2)

$y = \log_e x$

FIGURE 8.2

variable, the table is best computed as if the opposite were true. That is, choose the value of y and then compute x.

Example 2 Graph $y = \log_e x$. Use the table in Appendix 3. (See Figure 8.2.)

Figure 8.3 shows graphs of equations of the form $y = \log_a x$, where $a > 0$ and $a \neq 1$. The x intercept of all equations of the form $y = \log_a x$ is 1, and $x = 0$ is a vertical asymptote of each.

Logarithmic equations of the form $y = \log_a f(x)$, where $f(x) \neq x$ are not the type illustrated in Figure 8.3.

Example 3 Graph $y = \log_2 (x - 1)$. (See Figure 8.4.)

Rewriting, $2^y = x - 1$. Therefore $x = 2^y + 1$. The x intercept is 2, and there is a vertical asymptote at $x = 1$.

FIGURE 8.3

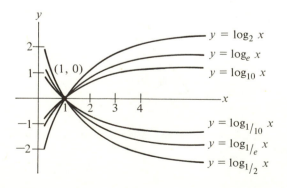

$y = \log_2 x$

$y = \log_e x$

$y = \log_{10} x$

(1, 0)

$y = \log_{1/10} x$

$y = \log_{1/e} x$

$y = \log_{1/2} x$

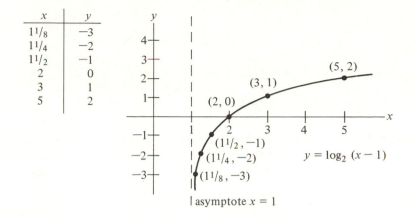

x	y
$1\frac{1}{8}$	-3
$1\frac{1}{4}$	-2
$1\frac{1}{2}$	-1
2	0
3	1
5	2

FIGURE 8.4

▶ **Exercise 8.1**

Graph each of the following.

1 $y = \log_3 x$ **2** $y = \log_5 x$

3 $y = \log_3 x^3$ **4** $y = \log_2 (x + 1)$

5 $y = \log_e (x - 1)$ **6** $y = \log_{10} x$

8.2 Differentiation of Logarithmic Functions

Differentiation of logarithmic functions can be considered in two cases. The first, simpler case is to differentiate logarithmic functions to the base e.

RULE 10

Let $y = \log_e f(x)$, where $f(x) > 0$; then

$$y' = \frac{f'(x)}{f(x)}$$

The derivative of the logarithm of an expression, base e, is the derivative of the expression divided by the expression.

Proof   If $y = \log_e f(x)$, then $e^y = f(x)$. Differentiating,

$$(y')(e^y) = f'(x)$$

Solving for y',

$$y' = \frac{f'(x)}{e^y}$$

Substituting for e^y,

$$y' = \frac{f'(x)}{f(x)}$$

Example 4   If $y = \log_e (x^3 + 2x + 3)$, find y'.

$$y' = \frac{(3x^2 + 2)}{x^3 + 2x + 3}$$

Example 5   If $y = \log_e (x^2 + 2x + 1)^6$, find y'.

There are two approaches to solving this problem. First, we can apply Rule 10 directly. Second, we can use the laws of logarithms to simplify the function.

First approach

$$y' = \frac{6(x^2 + 2x + 1)^5(2x + 2)}{(x^2 + 2x + 1)^6}$$

$$= \frac{6(2x + 2)}{x^2 + 2x + 1}$$

$$= \frac{12(x + 1)}{(x + 1)(x + 1)}$$

$$= \frac{12}{x + 1}$$

Second approach

$$y = \log_e (x^2 + 2x + 1)^6$$

By the third law of logarithms,

$$y = 6 \log_e (x^2 + 2x + 1)$$

Applying Rule 10,

$$y' = 6\left(\frac{2x + 2}{x^2 + 2x + 1}\right) = 6\left[\frac{2(x + 1)}{(x + 1)(x + 1)}\right]$$

$$= \frac{12}{x + 1}$$

Example 6   If $y = \log_e \sqrt{2x + 3}$, find y'.

First approach

$$y' = \frac{\frac{1}{2}(2x + 3)^{-1/2}(2)}{(2x + 3)^{1/2}} = \frac{1}{2x + 3}$$

Second approach

By the third law of logarithms,

$$y = \tfrac{1}{2} \log_e (2x + 3)$$

Applying Rule 10,

$$y' = \frac{1}{2}\left(\frac{2}{2x+3}\right) = \frac{1}{2x+3}$$

To differentiate a function of the type $y = \log_a f(x)$ we can use the change of base formula to rewrite the function using base e and then differentiate by Rule 10. However, Rule 11, of which Rule 10 is a special case, might simplify the computations.

RULE 11

If $y = \log_a f(x)$ and $f(x) > 0$, then

$$y' = \frac{f'(x)}{(\log_e a) f(x)} \quad (\log_e a \text{ is a constant})$$

The log of a quantity to base a is the derivative of the quantity over the quantity multiplied by $\log_e a$.

Proof   Let $y = \log_a f(x)$. Then by the change of base formula,

$$y = \frac{\log_e f(x)}{\log_e a}$$

Applying Rule 10,

$$y' = \frac{f'(x)}{(\log_e a)[f(x)]}$$

Example 7   If $y = \log_{10}(2x)$, find y'.

$$y' = \frac{2}{(\log_e 10)(2x)} = \frac{1}{(\log_e 10)(x)}$$

Example 8   If $y = \log_3 [(x-3)(x^2+6)]$, find y'.

$$y' = \frac{(x-3)(2x) + (x^2+6)(1)}{(\log_e 3)(x-3)(x^2+6)}$$

$$= \frac{3x^2 - 6x + 6}{(\log_e 3)(x-3)(x^2+6)}$$

▶ **Exercise 8.2**

Find y' in Problems 1 through 9.

  1 $y = \log_e x$

  2 $y = \log_e (x^2 + 3x - 4)$

3 $y = \log_e (x^3 - 3x)$

4 $y = \log_e (x + 1)^3$

5 $y = \log_e (2x + 5)^7$

6 $y = \log_e [(x + 1)(2x^2 - 3)]$

7 $y = \log_e \left(\dfrac{2x + 1}{x - 3} \right)$

8 $y = \log_e (x + 1) + \log_e (2x + 7)$

9 $y = \log_e e^{3x}$

10 A certain species of plant grows according to

$$\log_e \left[\frac{y}{105 - y} \right] = .1(x - 10)$$

where x is the age of the plant in years and y is the height of the plant in feet. What is the approximate height of the plant when it is 5 years old? Find the rate of growth of the plant. (*Hint:* Use the second law of logarithms, set the equation equal to zero, and use Rule 9 (Chapter 7).)

11 Does $y = \log_e x$ have a maximum or a minimum? Explain your answer.

12 A foreman has discovered that the number of units of work per day for the average employee is related to his increase in salary by $u = 10 + 2 \log_e (2R + 1)$, where R is his hourly pay raise and u is the units of work per day.

a What is the rate of change in work output u with respect to R?

b What is the rate of change in output for a $1.50 raise?

c How many units of work will the average employee produce with no raise?

Find y' in Problems 13 through 20.

13 $y = \log_{10} x$

14 $y = \log_{10} (3x - 1)$

15 $y = \log_{10} (2x^2 - 6x)$

16 $y = \log_5 (x^2 - 3x + 4)$

17 $y = \log_{10} (x^3 - 6)$

18 $y = \log_2 (2x - 2)^5$

19 $y = \log_4 (x + 3)$

20 $y = \log_3 e^{3x}$

21 What is the slope of the tangent line to the graph of $y = \log_{10} x$ at $x = 2$?

22 The derivative of $y = \log_a f(x)$ is sometimes given as $y' = (\log_a e)[f'(x)]/f(x)$. Justify this formula.

8.3 Differentiation of a Function to a Functional Exponent

The techniques developed so far do not allow us to differentiate functions such as $y = x^x$, $y = (2x + 1)^{x-5}$, and so on, where both the base and the exponent are nonconstant functions.

RULE 12

Let $y = g(x)^{h(x)}$; then

$$y' = g(x)^{h(x)}\left[\frac{h(x)\,g'(x)}{g(x)} + h'(x)\log_e g(x)\right]$$

Example 9   If $y = x^x$, find y'.

$$y' = x^x\left[\frac{x(1)}{x} + 1\log_e x\right] = x^x(1 + \log_e x)$$

Although Rule 12 can be applied automatically, it is cumbersome and difficult to memorize, so we will examine another technique.

Every problem of the type $y = g(x)^{h(x)}$ can be differentiated by using the following procedure:

1 Take the logarithm to the base e of both sides of the equation to be differentiated.

2 Simplify the resulting equation by applying the laws of logarithms.

3 Differentiate both sides of the equation.

4 Solve the resulting equation for y' and substitute for y.

Example 10   If $y = 2^x$, find y'.

$$1. \ \log_e y = \log_e 2^x$$

$$2. \ \log_e y = x\log_e 2$$

$$3. \quad \frac{y'}{y} = x(0) + \log_e 2$$

$$4. \quad y' = y\log_e 2 = 2^x\log_e 2$$

Example 11   If $y = e^x$, find y'.

$$1. \ \log_e y = \log_e e^x$$

$$2. \ \log_e y = x\log_e e$$

$$3. \quad \frac{y'}{y} = 1$$

$$4. \quad y' = y = e^x$$

▶ **Exercise 8.3**

Find y' in Problems 1 through 6. Use Rule 12.

1 $y = x^{2x}$ **2** $y = (x + 1)^x$

3 $y = xe^x$

4 $y = (x+1)^{2x}$

5 $y = (e^x)e^x$

6 $y = (x+2)^{x+3}$

Find y' in Problems 7 through 9, using the four-step procedure given in this section.

7 $y = x^{3x}$

8 $y = 3^x$

9 $y = 3^{e^x}$

10 Prove Rule 12.

8.4 Integration Yielding Logarithmic Functions

Rule 10 for differentiation can be used in reverse to integrate functions of the type $f'(x)/f(x)$. This is stated in the following integration formula.

FORMULA 9

If $y = f'(x)/f(x)$, then

$$\int y\, dx = \int \frac{f'(x)}{f(x)}\, dx = \log_e f(x) + c$$

where $f'(x) \neq 0$ and $f(x) > 0$.

Example 12 Find $\int (1/x)\, dx$, where $x > 0$.

Let $f(x) = x$; then $f'(x) = 1$. Therefore $1/x$ is of the form $f'(x)/f(x)$. Applying Formula 9,

$$\int \left(\frac{1}{x}\right) dx = \log_e x + c$$

Note that $1/x = x^{-1}$, but Formula 3 does not apply. If the exponent is increased by 1, the new exponent is zero and $x^0/0$ is meaningless. This is why Formula 3 was restricted to an exponent other than -1.

Example 13 Find $\int [2x/(x^2+1)]\, dx$.

Consider $f(x) = x^2 + 1$; then $f'(x) = 2x$. Therefore Formula 9 applies, and we have

$$\int \frac{2x}{x^2+1}\, dx = \log_e (x^2+1) + c$$

Example 14 Find $\int [x^2/(x^3-6)]\, dx$, where $x^3 - 6 > 0$.

Let $f(x) = x^3 - 6$; then $f'(x) = 3x^2$. Note that $x^2/(x^3 - 6)$ is not of the form $f'(x)/f(x)$, because the numerator does not contain the factor 3. However, this can be remedied by rewriting as follows.

$$\int \frac{x^2}{x^3 - 6} \, dx = \tfrac{1}{3} \int \frac{3x^2}{x^3 - 6} \, dx$$

Formula 9 can now be applied, yielding

$$\tfrac{1}{3} \int \frac{3x^2}{x^3 - 6} \, dx = \tfrac{1}{3} \log_e (x^3 - 6) + c$$

▶ **Exercise 8.4**

Integrate each of the following.

1 $\int \dfrac{2x}{x^2 + 5} \, dx$ 2 $\int \dfrac{3x^2}{x^3 - 10} \, dx$

3 $\int \dfrac{4x + 1}{2x^2 + x + 1} \, dx$ 4 $\int \dfrac{e^x}{e^x - 1} \, dx$

5 $\int \dfrac{2}{2x + 3} \, dx$ 6 $\int \dfrac{6}{2x + 3} \, dx$

7 $\int \dfrac{t^2}{2t^3 + 3} \, dt$ 8 $\int \dfrac{1}{x + 1} \, dx$

9 $\int \dfrac{(\log_e x)^5}{x} \, dx$

8.5   Differential Equations Revisited

In many instances in the fields of biology, business, economics, and so on, the rate of change dQ/dt is directly proportional to the amount Q of some quantity present at time t. That is, $dQ/dt = kQ$, where k is some constant.

Applied Example 15   The rate of change of population p with respect to time t is directly proportional to the number of people present at any time. This can be represented by the differential equation $dp/dt = kp$, where k is a constant. To find the equation relating population to time we solve $dp/dt = kp$.

Rewriting and separating the variables, $dp/p = k \, dt$. Integrating both sides yields $\log_e p = kt + c$. Hence, $p = e^{kt+c}$.

In certain situations we can determine two corresponding values of the variables. This allows us to determine the values of the constants.

Applied Example 15 (Continued)   At the time of the last census a rapidly growing island nation had a population of 2981. Find the equation relating population to time.

Consider the census to have been at $t = 0$. Then $2981 = e^{k(0)+c}$. Solving,

$2981 = e^c$. Therefore, by the table in Appendix 3, $c = 8$. Thus the population is $p = e^{kt+8}$.

One year later, the population had increased to 3290. So

$$3290 = e^{k+8}$$
$$3290 = e^k e^8$$
$$\frac{3290}{e^8} = e^k$$
$$1.104 \doteq e^k$$

By Appendix 3, $k \doteq .1$. Therefore the population function is $p = e^{.1t+8}$.

What is the population 10 years after the census?

$$p = e^{.1(10)+8} = e^9 \doteq 8013$$

▶ **Exercise 8.5**

1 Solve $dx/x = 2\ dt$.

2 Solve $dv/v = .2\ dt$.

3 Solve $dw/w = a\ dt$, where a is a constant.

In Problems 4 through 6 find the differential equation. Find a solution of the differential equation.

4 The rate of radioactive disintegration of an element with respect to time is directly proportional to the amount of the element present at any time.

5 The rate of growth of a bacteria colony with respect to time is directly proportional to the number of bacteria present at any time.

6 An electric capacitor discharges at a rate directly proportional to the charge remaining.

7 If $100 is invested at 6% compounded continuously, what will the amount be at the end of 5 years? (Assume that the rate of growth of the principal is directly proportional to the amount of money present at any time and that $k = .06$.)

8 In Problem 5, if the number of bacteria increases from 100 to 200 as t goes from 1 to 2, find the value of the constants c and k.

8.6 Applications

Applied Example 16 A boat manufacturer determined after careful study that his marginal cost Q' of producing a certain model of sailboat is given by

$$Q' = \frac{600}{.02x + 5.4}$$

where x is the number of boats produced in a normal production run. Find the cost for producing x sailboats if the fixed cost is \$1000.

$$\text{Total cost } Q = \int Q' \, dx = \int \frac{600}{.02x + 5.4} \, dx$$

$$= \frac{600}{.02} \int \frac{.02}{.02x + 5.4} \, dx$$

$$= \frac{600}{.02} [\log_e (.02x + 5.4)] + c$$

If the fixed cost is $Q = 1000$ when $x = 0$,

$$c = 1000(1 - 30 \log_e 5.4) \doteq -50,000$$

Therefore

$$Q \doteq 30,000 \log_e (.02x + 5.4) - 50,000$$

Find the total cost of producing 100 sailboats.

$$Q \doteq 30,000 \log_e (.02(100) + 5.4) - 50,000$$
$$\doteq 30,000 \log_e (7.4) - 50,000$$
$$\doteq 30,000(2) - 50,000$$
$$\doteq \$10,000$$

If the manufacturer wants to make a 20% profit on each boat, what is the selling price of each boat? Since each boat costs \$100, the profit is 20% of \$100, or \$20. Therefore each boat must sell for \$120.

Applied Example 17 When an electromotive force is applied to an electric circuit, the current begins rising toward its full value. The full value of the current is not reached immediately, because a magnetic field is formed around the circuit first. The strength of the magnetic field is the inductance of the circuit. For a circuit containing resistance and inductance we have at any instant after the switch is closed the relation

$$E = Ri + L\frac{di}{dt}$$

where t is the time elapsed after the switch is closed, R is the resistance, i is the current, E is the electromotive force, and L is the inductance of the circuit. To find the current i at any time t we rewrite the equation as

$$\frac{-R}{E - Ri} \, di = -\frac{R}{L} \, dt$$

and integrate.

Since the left member is of the form $f'(x)/f(x)$, we have

$$\log_e (E - Ri) = \int -\frac{R}{L} \, dt$$

In any given circuit R and L are constant; hence the equation may be rewritten

$$\log_e (E - Ri) = -\frac{R}{L} \int dt$$

Therefore

$$\log_e (E - Ri) = -\frac{Rt}{L} + c$$

At the time $t = 0$, $i = 0$, so $c = \log_e E$. Therefore

$$\log_e (E - Ri) = -\frac{Rt}{L} + \log_e E$$

Rearranging the members of the equation,

$$\log_e (E - Ri) - \log_e E = -\frac{Rt}{L}$$

By the second law of logarithms,

$$\log_e \frac{E - Ri}{E} = -\frac{Rt}{L}$$

Hence

$$e^{-Rt/L} = \frac{E - Ri}{E}$$

$$e^{-Rt/L} = 1 - \frac{Ri}{E}$$

$$e^{-Rt/L} - 1 = -\frac{Ri}{E}$$

$$1 - e^{-Rt/L} = \frac{Ri}{E}$$

$$\frac{E}{R}(1 - e^{-Rt/L}) = i$$

Note that this is a learning curve. As t increases, i approaches E/R. (See Figure 8.5.)

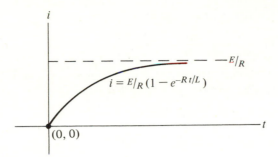

FIGURE 8.5

▶ **Exercise 8.6**

1 A theater owner has determined that his marginal profit is given by $P' = 50/(x + 1)$, where x is the number of patrons attending each showing. Find the total profit function for any showing if the fixed cost is 100. If 402 patrons attend a single showing, what is the total profit?

Evaluate the definite integrals in Problems 2 through 4.

2 $\displaystyle\int_0^1 \frac{x}{x^2 + 1}\, dx$ **3** $\displaystyle\int_2^3 \frac{1}{x}\, dx$ **4** $\displaystyle\int_0^2 \frac{1}{x + 3}\, dx$

Evaluate the improper integrals in Problems 5 through 7.

5 $\displaystyle\int_1^\infty \frac{1}{x}\, dx$ **6** $\displaystyle\int_0^\infty \frac{1}{x + 1}\, dx$ **7** $\displaystyle\int_0^1 \frac{1}{x}\, dx$

8 Show that $y = \log_e x$ is always increasing.

9 Show that the area bounded by the lines $x = 1$, $x = 3$, and the curve $y = 2/(2x + 3)$ is $\log_e 1.8$.

10 Show that the function $y = \log_e 2x$ is concave downwards for its domain $x > 0$.

11 Show that $y = \log_e e^{x^2 - 4}$ has a minimum at $(0, -4)$.

12 Graph $y = \log_e 10^x$.

13 Find the rate of change of t with respect to x in the equation

$$t = \log_e \left(\frac{3}{3 - x} \right)$$

14 Find the derivative of

$$y = \log_e \left[\frac{\sqrt{x^2 + 1}}{(x + 1)^3} \right]$$

15 Water is filling a reservoir at a rate given by

$$\frac{dv}{dt} = \frac{1000 \text{ gal/hr}}{1+t}$$

where t is the time in hours and v is the volume in gallons. Find the volume of water that flowed into the tank from $t = 0$ to $t = 4$ hr.

16 Find the inflection point of the function $y = e^{-1/x}$.

17 Find the first partials of the following equations:

    a $z = \log_e (x + y)^2$ **b** $z = x^2 \log_e (x^2 + y)$

---

self-test · chapter eight

Graph the equations in Problems 1 and 2.

1 $y = \log_5 x$ **2** $y = \log_2 (x + 3)$

Differentiate the equations in Problems 3 through 10.

3 $y = \log_e 3x$ **4** $y = \log_e e^{x^2}$

5 $y = \log_e \left(\dfrac{x}{x+1} \right)$ **6** $y = \log_e \sqrt[3]{x+2}$

7 $y = \log_2 3x$ **8** $y = \log_{10} e^{x^2}$

9 $y = x^{5x}$ **10** $y = 4^{x-1}$

Integrate the expressions in Problems 11 through 13.

11 $\displaystyle\int \frac{3}{x+1}\, dx$ **12** $\displaystyle\int \frac{3x}{\frac{3}{2}x^2 + 5}\, dx$ **13** $\displaystyle\int \frac{dx}{2x+1}$

Solve the differential equations in Problems 14 and 15.

14 $\dfrac{dQ}{dt} = 2Q$ **15** $\dfrac{dy}{dx} = xy$

16 In a tank leaking at the bottom the rate of change of the height of the water with respect to time is 5% of the height remaining at any time. Find the equation of the height of the water at any time if $h = 20.086$ when $t = 0$.

17 The rate of change of cost C (in dollars) of a rare gem with respect to a change in weight w (in carats) is $1800w$. Find the cost of a 2-carat gem.

appendix one

review of exponents

The topic of exponents is so vital to the study of calculus that it seems appropriate to review the basic definitions and laws of exponents.

In an expression of the form x^n, x is called the *base* and n is called the *exponent*.

Example 1

In x^3, x is the base and 3 is the exponent.

In $5t^4$, t is the base and 4 is the exponent.

In $(5t)^4$, $5t$ is the base and 4 is the exponent.

The Laws of Exponents

THE FIRST LAW OF EXPONENTS

$$a^m \cdot a^n = a^{m+n}$$

where n and m are rational numbers and $a > 0$.

Example 2

$$x^3 \cdot x^4 = x^{4+3} = x^7$$

$$2x^7 \cdot 3x^2 = 6x^9$$

$$2(x + 5) \cdot 3(x + 5)^3 = 6(x + 5)^4$$

THE SECOND LAW OF EXPONENTS

$$\frac{a^m}{a^n} = a^{m-n}$$

where m and n are rational numbers and $a > 0$.

Example 3

$$\frac{a^6}{a^3} = a^{6-3} = a^3$$

$$\frac{12x^7}{4x^3} = 3x^4$$

$$\frac{x^{y+3}}{x^{y-2}} = x^{(y+3)-(y-2)} = x^5$$

THE THIRD LAW OF EXPONENTS

$$(a^n)^m = a^{nm}$$

where m and n are rational numbers and $a > 0$.

Example 4

$$(x^2)^3 = x^6$$

$$(e^x)^2 = e^{2x}$$

$$[(3+x)^3]^2 = (3+x)^6$$

▶ Exercise 1

Use the three laws of exponents to perform the indicated operations.

1 $x^6 \cdot x^4$

2 $y^a \cdot y$

3 $\dfrac{(x+3)^5}{(x+3)^2}$

4 $(x^3)^4$

5 $y^{a+2} \cdot y^{a-3}$

6 $(e^x)^3$

7 $(x+1)^y(x+1)^a(x+1)^3$

8 $\dfrac{10^7}{10^3}$

9 $[(x+2)^2]^5$

Definitions of Zero and Negative Exponents

DEFINITION

An exponent that is a counting number denotes the number of times that the base is used as a factor.

Example 5

$$2^5 \text{ means } 2 \cdot 2 \cdot 2 \cdot 2 \cdot 2 = 32$$
$$x^4 \text{ means } x \cdot x \cdot x \cdot x$$
$$x^1 \text{ means } x$$

DEFINITION

$a^0 = 1$, where a is a real number and $a \neq 0$.

Example 6

$$x^0 = 1$$
$$(x + 3)^0 = 1$$
$$(5)^0 = 1$$

DEFINITION

$a^{-n} = 1/a^n$, where a is a real number and $a \neq 0$.

Example 7

$$x^{-1} = \frac{1}{x}$$

$$2x^{-3} = \frac{2}{x^3}$$

$$\frac{4}{x^{-6}} = 4x^6$$

▶ **Exercise 2**

In Problems 1 through 6 use the definitions to make all exponents positive. Use the three laws of exponents when they apply.

1 2^0 **2** x^{-3} **3** $2x^{-5}$

4 $\dfrac{4a^{-3}}{b^{-2}}$ **5** $\dfrac{x^7}{x^{-3}}$ **6** $(y^{-3})^2$

Rewrite Problems 7 through 12, making all exponents of literal expressions negative.

7 $\dfrac{1}{x}$ **8** $\dfrac{3}{x^2}$ **9** $\dfrac{1}{x + 2}$

10 $\dfrac{1}{xy}$ **11** x^3 **12** $\dfrac{9}{(x+3)^2}$

Definition of Fractional Exponents

DEFINITION

$$x^{n/m} = (x^{1/m})^n = \sqrt[m]{x^n} = \left(\sqrt[m]{x}\right)^n$$

where m is a counting number greater than 1 and n is an integer. (x must be nonnegative when m is an even counting number and n/m is reduced to lowest terms.)

Example 8   Rewrite each of the following in radical form. Evaluate whenever possible.

$$8^{2/3} = \sqrt[3]{8^2} = \sqrt[3]{64} = 4$$

$$16^{-3/4} = \frac{1}{16^{3/4}} = \frac{1}{\left(\sqrt[4]{16}\right)^3} = \frac{1}{2^3} = \frac{1}{8}$$

$$x^{2/3} = \sqrt[3]{x^2} = \left(\sqrt[3]{x}\right)^2$$

Example 9   Rewrite each of the following in exponential form.

$$\sqrt[4]{x^3} = x^{3/4}$$

$$\frac{2}{\sqrt{x}} = \frac{2}{x^{1/2}} = 2x^{-1/2}$$

$$\sqrt[3]{x+3} = (x+3)^{1/3}$$

▶ Exercise 3

Write Problems 1 through 6 in radical form. Evaluate whenever possible.

1 $25^{1/2}$ **2** $25^{-1/2}$ **3** $81^{3/4}$

4 $9^{-3/2}$ **5** $64^{1/2}/8^{1/3}$ **6** $(x^{2/5})^2$

Write Problems 7 through 12 in exponential form, with the exponent in the numerator of the expression.

7 $\sqrt[3]{x}$ **8** $\sqrt{x^2+4}$ **9** $\dfrac{1}{\sqrt{x}}$

10 $\dfrac{3}{\sqrt[3]{x+5}}$ **11** $\dfrac{x}{\sqrt{x^2-4}}$ **12** $x^2\sqrt{x^3-9}$

self-test · appendix one

Identify the base and exponent in Problems 1 through 3.

1 x^4 **2** $6y^2$ **3** $(6y)^2$

Use the three laws of exponents to perform the operations indicated in Problems 4 through 6.

4 $x^5 \cdot x \cdot x^b$ **5** $\dfrac{x^{a+7}}{x^{a+3}}$ **6** $(x^{a-3})^2$

Rewrite Problems 7 through 9, making all exponents positive.

7 x^{-3} **8** $3x^{-4}$ **9** $\dfrac{1}{x^{-7}}$

Rewrite Problems 10 through 12, making all exponents negative.

10 $\dfrac{1}{x^{1/2}}$ **11** x^5 **12** $\dfrac{1}{x-1}$

Rewrite Problems 13 through 15 in radical form.

13 $x^{5/3}$ **14** $a^{-1/4}$, where $a > 0$ **15** $3(x-1)^{1/2}$

Rewrite Problems 16 through 18 in exponential form.

16 $\sqrt[5]{x}$ **17** $\sqrt{x^2 + 25}$ **18** $\dfrac{2}{\sqrt[3]{c+7}}$

introduction to logarithms

This appendix is a basic discussion of the concept of a logarithm.

In the equation $3^2 = 9$, 2 is called the *exponent* of the base 3. However the number 2 is also closely related to the number 9 in the equation. What is the name of the relation of 2 to 9? Two is the *logarithm* of 9 when the base is 3. Therefore the number 2 in the equation $3^2 = 9$ has two names. It is called the *exponent* of base 3 and the *logarithm* of 9 when the base is 3.

Example 1   Let $2^3 = 8$.

Three is the exponent of 2. Three is the logarithm of 8 when the base is 2.

Example 2   Let $27^{2/3} = 9$.

Two-thirds is the exponent of 27. Two-thirds is the logarithm of 9 when the base is 27.

Example 3   Let $e^2 \doteq 7.389$.

Two is the exponent of e. Two is the logarithm of 7.389 when the base is e.

The notation for the logarithm of a number is as follows: If $3^2 = 9$, then 2 is the logarithm of 9 when the base is 3. This is written symbolically

$$2 = \log_3 9$$

which is read, "2 is the logarithm of 9, base 3."

Example 4

$$\text{If } 2^3 = 8, \text{ then } 3 = \log_2 8 \text{ (or } \log_2 8 = 3).$$
$$\text{If } 4^{1/2} = 2, \text{ then } \tfrac{1}{2} = \log_4 2 \text{ (or } \log_4 2 = \tfrac{1}{2}).$$

If $10^{-2} = 1/100$, then $-2 = \log_{10}(1/100)$.

If $e^0 = 1$, then $0 = \log_e 1$.

▶ **Exercise 1**

For Problems 1 through 9, fill in the blank with the correct word.

1 Let $4^3 = 64$.

Three is the _____ of 4.

Three is the _____ of 64 when the base is _____.

2 Let $16^{1/4} = 2$.

One-fourth is the _____ of 16.

One-fourth is the _____ of 2 when the _____ is 16.

3 Let $e^3 \doteq 20.086$.

Three is the _____ of e.

Three is the _____ of 20.086 when the _____ is

_____.

4 Let $8^{2/3} = 4$.

Two-thirds is the _____ of 8.

Two-thirds is the _____ of _____ when the _____

is _____.

5 Let $e^{.1} \doteq 1.105$.

One-tenth is the _____ of _____.

One-tenth is the _____ of _____ when the _____

is _____.

6 Let $4^0 = 1$.

_____ is the exponent of 4.

Zero is the _____ of 1 when the base is _____.

7 Let $100^{1/2} = $ _____.

One-half is the _____ of _____.

One-half is the logarithm of _____ when the base is _____.

8 Is zero always the logarithm of 1? Explain.

9 Is there a number that is the logarithm of 0? Explain.

Express Problems 10 through 18 using logarithm notation.

10 $2^4 = 16$ **11** $4^3 = 64$ **12** $5^2 = 25$

13 $10^3 = 1000$ **14** $10^{-1} = .1$ **15** $e^{1/2} \doteq 1.649$

16 $e^{-2} \doteq .1353$ **17** $16^{-1/4} = \frac{1}{2}$ **18** $12^0 = 1$

Write the logarithmic expressions in Problems 19 through 24 in exponential notation.

19 $5 = \log_2 32$

20 $\frac{1}{2} = \log_{25} 5$

21 $-1 = \log_{10} .1$

22 $\log_e 12.182 = 2.5$

23 $\log_9 \frac{1}{81} = -2$

24 $\log_{10} 1000 = 3$

The Definition of a Logarithm

DEFINITION

If $b^a = N$, then $a = \log_b N$ provided $b > 0$, $b \neq 1$, and $N > 0$.

The number N is positive, because a positive base b raised to any exponent is a positive quantity. Therefore the $\log_b N$ where $N \leq 0$ is a meaningless expression. For all future logarithmic expressions, the quantity in the N position is assumed to be positive even if it is not so stated.

Although the definition allows us to use all positive numbers except 1 as the base for logarithmic expressions, the most common bases are e and 10. Logarithms to the base e are called *Naperian logarithms* or *natural logarithms* after the inventor of logarithms, John Napier. Logarithms to the base 10 are called *Briggs's logarithms* or *common logarithms* after Henry Briggs, who collaborated with John Napier in computing the first table of logarithms to the base 10.

THEOREM

If $\log_b M = \log_b N$, then $M = N$.

If two quantities M and N have the same logarithm to the same base, then M and N are equal.

Proof   Let $\log_b M = a$; then $b^a = M$. Also $\log_b N = a$; then $b^a = N$. Therefore $M = b^a = N$, so $M = N$.

Example 5   If $\log_2 x = \log_2 6$, then $x = 6$.

Example 6   Solve the following equation for x: $\log_2 (3x + 1) = 2$.

Using the definition of a logarithm, we may rewrite the equation as

$$3x + 1 = 2^2$$
$$3x + 1 = 4$$
$$3x = 3$$
$$x = 1$$

▶ **Exercise 2**

In Problems 1 through 8 compute the value of x.

1 $\log_2 8 = x$ **2** $\log_e x = 2$

3 $\log_{10} x = 2$ **4** $\log_{32} 2 = x$

5 $\log_x 27 = 3$ **6** $\log_2 (x + 2) = 3$

7 $\log_3 (x - 1) = 1$ **8** $\log_2 (x^2 - x - 5) = 0$

9 The mean life m of a radioactive substance is related to the half-life h as given by $m = h/\log_e 2$. If the half-life of krypton is 2.4 sec, what is the mean life?

Solve for x in Problems 10 through 12.

10 $\log_3 x = \log_3 11$

11 $\log_6 (x + 1) = \log_6 5$

12 $\log_b x^3 = \log_b 8$

The Laws of Logarithms

Often we want to determine the logarithm of a quantity that is a product, quotient, or power. Logarithms of such quantities may be found by applying the three laws of logarithms. The three laws of logarithms correspond to the three laws of exponents. Each of the laws of logarithms will be proved using the laws of exponents.

THE FIRST LAW OF LOGARITHMS

$$\log_b mn = \log_b m + \log_b n$$

The logarithm of a product of two numbers is the logarithm of the first factor added to the logarithm of the second factor, provided the bases are the same.

Proof We begin by observing that if $b^a = m$ and $b^c = n$, then $a = \log_b m$ and $c = \log_b n$. Then

$$mn = b^a \cdot b^c = b^{a+c} = b^{\log_b m + \log_b n}$$

Therefore $\log_b mn = \log_b m + \log_b n$. This proves the first law of logarithms.

Example 7 Find $\log_2 (8)(32)$ using the first law of logarithms.

$$\log_2 (8)(32) = \log_2 8 + \log_2 32 = 3 + 5 = 8$$

THE SECOND LAW OF LOGARITHMS

$$\log_b\left(\frac{m}{n}\right) = \log_b m - \log_b n$$

The logarithm of a quotient is the logarithm of the numerator minus the logarithm of the denominator, provided the bases are equal.

Proof   Let $b^a = m$ and $b^c = n$; then $a = \log_b m$ and $c = \log_b n$. Then

$$\frac{m}{n} = \frac{b^a}{b^c} = b^{a-c} = b^{\log_b m - \log_b n}$$

Therefore $\log_b (m/n) = \log_b m - \log_b n$. This proves the second law of logarithms.

Example 8

$$\log_2\left(\frac{32}{8}\right) = \log_2 32 - \log_2 8 = 5 - 3 = 2$$

THE THIRD LAW OF LOGARITHMS

$$\log_b m^c = c \cdot \log_b m$$

The logarithm of a number to an exponent is the exponent multiplied by the logarithm of the number.

Proof   Let $b^a = m$; then $a = \log_b m$. Since $m^c = (b^a)^c$, we have $m^c = b^{c \cdot \log_b m}$
Therefore $\log_b m^c = c \log_b m$. This proves the third law of logarithms.

Example 9

$$\log_2 4^3 = 3 \log_2 4 = 3(2) = 6$$
$$\log_3 9^5 = 5 \log_3 9 = 5(2) = 10$$

▶ Exercise 3

Find the logarithms of the products indicated in Problems 1 through 6 by using the first law of logarithms.

1 $\log_2 (16)(32)$ **2** $\log_3 (27)(3)$ **3** $\log_4 (16)(64)$

4 $\log_8 (2)(64)$ **5** $\log_a xy$ **6** $\log_7 (49)(7)$

Write Problems 7 through 9 as the logarithms of products by using the first law of logarithms in reverse.

7 $\log_a x + \log_a y$ **8** $\log_2 3 + \log_2 9$ **9** $\log_6 100 + \log_6 x^2$

Find the logarithms of the quotients in Problems 10 through 15.

10 $\log_{36}\left(\dfrac{216}{36}\right)$
$\qquad$ **11** $\log_2\left(\dfrac{128}{32}\right)$
$\qquad$ **12** $\log_3\left(\dfrac{9}{27}\right)$

13 $\log_5\left(\dfrac{1}{125}\right)$
$\qquad$ **14** $\log_a\left(\dfrac{x}{y}\right)$
$\qquad$ **15** $\log_4\left(\dfrac{64}{4}\right)$

Write Problems 16 through 18 as the logarithms of quotients.

16 $\log_a x - \log_a y$
$\qquad$ **17** $\log_5 27 - \log_5 10$
$\qquad$ **18** $\log_8 100 - \log_8 x^3$

Find the logarithms of Problems 19 through 21 by using the third law of logarithms.

19 $\log_2 4^2$
$\qquad$ **20** $\log_5 25^3$
$\qquad$ **21** $\log_{10} 1000^2$

Express Problems 22 through 24 as the logarithms of numbers to an exponent.

22 $2 \log_5 5$
$\qquad$ **23** $a \log_b x$
$\qquad$ **24** $2 \log_5 6$

Solve for x in Problems 25 through 27.

25 $2 \log_2 x = 2$
$\qquad$ **26** $\frac{1}{2} \log_x 16 = 2$
$\qquad$ **27** $2 \log_4 (x - 1) = 1$

Summary

The three laws of logarithms may be used in combination.

Example 10 Express $\log_2 x^2 y/z$ as the logarithms of individual expressions.

$$\log_2\left(\frac{x^2 y}{z}\right) = \log_2 x^2 y - \log_2 z$$

$$= \log_2 x^2 + \log_2 y - \log_2 z$$

$$= 2 \log_2 x + \log_2 y - \log_2 z$$

Example 11 Express the following as the logarithm of a single expression:
$3 \log_3 a + \log_3 b - \log_3 c$.

$$3 \log_3 a + \log_3 b - \log_3 c = \log_3 a^3 + \log_3 b - \log_3 c$$

$$= \log_3 a^3 b - \log_3 c$$

$$= \log_3\left(\frac{a^3 b}{c}\right)$$

Example 12 Solve for x in the expression $\log_3 (x + 2) = 2$.

$$\log_3 (x + 2) = 2$$

$$x + 2 = 3^2$$

$$x + 2 = 9$$

$$x = 7$$

▶ **Exercise 4**

Express Problems 1 through 3 as the logarithms of individual expressions.

1 $\log_a xy^2$ **2** $\log_a \left(\dfrac{x^2 y^3}{z} \right)$ **3** $\log_a (x^4 y)$

Express Problems 4 and 5 as the logarithms of single expressions.

4 $3 \log_a x - \log_a y$ **5** $\log_a x + 2 \log_a y - 3 \log_a z$

Solve for x in Problems 6 through 9.

6 $\log_3 (x - 5) = \log_3 4$ **7** $\log_2 (x + 3) = 1$

8 $\log_3 (x - 5) = 2$ **9** $\log_a 2 + \log_a (x + 3) = \log_a 5$

Change of Base Formula

Suppose we know the $\log_a n$ and the $\log_a b$, can we find $\log_b n$. Yes. All that is needed is the change of base formula.

THE CHANGE OF BASE FORMULA

$$\log_b n = \frac{\log_a n}{\log_a b}$$

Example 13 If $\log_4 16 = 2$ and $\log_4 32 = \frac{5}{2}$, find the $\log_{32} 16$.
  By the change of base formula,

$$\log_{32} 16 = \frac{\log_4 16}{\log_4 32} = \frac{2}{\frac{5}{2}} = \frac{4}{5}$$

Proof Let $\log_b n = w$; then $b^w = n$. To change the base let $q = \log_a b$; then $b = a^q$. Substituting for b, $(a^q)^w = n$, and $a^{qw} = n$. Writing in logarithmic notation, $qw = \log_a n$. Hence, $\log_a b \cdot \log_b n = \log_a n$. Solving, $\log_b n = \log_a n / \log_a b$.

The change of base formula can be used to rewrite logarithmic equations in terms of different bases.

Example 14 Express $y = \log_{10} (3x^2 + 1)$ in base e.
  By the change of base formula,

$$y = \frac{\log_e (3x^2 + 1)}{\log_e 10}$$

Example 15 Show that $\log_e 10 = 1/\log_{10} e$.
  By the change of base formula,

$$\log_e 10 = \frac{\log_{10} 10}{\log_{10} e} = \frac{1}{\log_{10} e}$$

▶ **Exercise 5**

1 Given $\log_2 8 = 3$ and $\log_2 16 = 4$, find $\log_{16} 8$.
2 Given $\log_3 27 = 3$ and $\log_3 9 = 2$, find $\log_9 27$.
3 Given $\log_5 25 = 2$ and $\log_5 125 = 3$, find $\log_{25} 125$.
4 Given $\log_{10} 36 = 1.5563$ and $\log_{10} 6 = .7782$, find $\log_6 36$.

Write the equations in Problems 5 through 8 in base e.

5 $y = \log_{10} x^3$
6 $y = \log_6 (3x + 1)$
7 $y = \log_{10} (2x + 4)$
8 $y = \log_{10} (x + 3)^5$
9 Prove that $\log_a b = 1/\log_b a$.

self-test · appendix two

1 Let $8^{-2/3} = \frac{1}{4}$. Negative two-thirds is the _____ of $\frac{1}{4}$ when the base is _____. Negative two-thirds is the _____ of 8.

2 Express $5^2 = 25$ in logarithmic notation.

3 Express $\log_9 (1/81) = -2$ in exponential notation.

4 Solve for x in the following equations.
   a $\log_2 x = -3$
   b $\log_x 3 = \frac{1}{2}$
   c $\log_2 \frac{1}{2} = x$
   d $\log_2 (x + 4) = 1$
   e $\log_5 x = \log_5 2$

5 Express $\log_b (3x^2/y)$ as the logarithm of individual expressions.

6 Express $\log_7 (50/73)$ as the difference of logarithms.

7 Express $\log_b x + \log_b y - \log_b z$ as the logarithm of a single expression.

8 Solve for x in the following equations.
   a $\log_3 (x + 5) = 2$
   b $\log_a 5 + \log_a (x - 2) = \log_a 20$

9 Given $\log_{10} 50 = 1.6990$ and $\log_{10} e = .4343$, find $\log_e 50$.

10 Write $y = \log_{10} x$ in base e.

appendix three

table of exponentials

x	e^x	e^{-x}	x	e^x	e^{-x}
0	1	1	2.8	16.44	.0608
.1	1.105	.9048	2.9	18.17	.0550
.2	1.221	.8187	3.0	20.09	.0498
.3	1.350	.7408	3.1	22.20	.0450
.4	1.492	.6703	3.2	24.53	.0408
.5	1.649	.6065	3.3	27.11	.0369
.6	1.822	.5488	3.4	29.96	.0334
.7	2.014	.4966	3.5	33.12	.0302
.8	2.226	.4493	3.6	36.60	.0273
.9	2.460	.4066	3.7	40.45	.0247
1.0	2.718	.3679	3.8	44.70	.0224
1.1	3.004	.3329	3.9	49.40	.0202
1.2	3.320	.3012	4.0	54.60	.0183
1.3	3.669	.2725	4.1	60.34	.0166
1.4	4.055	.2466	4.2	66.69	.0150
1.5	4.482	.2231	4.3	73.70	.0136
1.6	4.953	.2019	4.4	81.45	.0123
1.7	5.474	.1827	4.5	90.02	.0111
1.8	6.050	.1653	4.6	99.48	.0101
1.9	6.686	.1496	4.7	109.95	.0091
2.0	7.389	.1353	4.8	121.51	.0082
2.1	8.166	.1225	4.9	134.29	.0074
2.2	9.025	.1108	5.0	148.41	.0067
2.3	9.974	.1003	6.0	403.43	.00248
2.4	11.02	.0907	7.0	1096.64	.000912
2.5	12.18	.0821	8.0	2980.98	.000335
2.6	13.46	.0743	9.0	8013.14	.0001234
2.7	14.88	.0672	10.0	22,026.65	.0000454

▶ **Exercise 1.1**

1

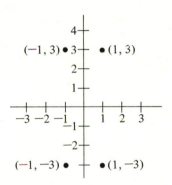

2

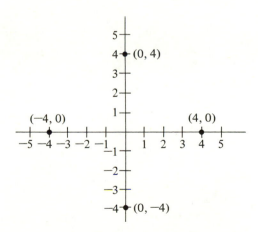

4

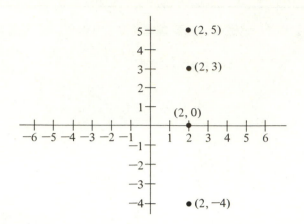

5

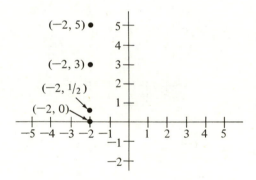

7

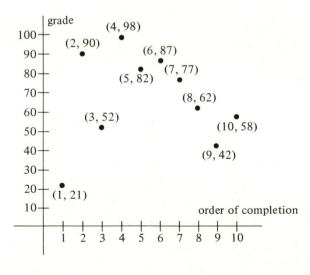

8

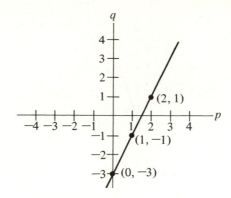

10

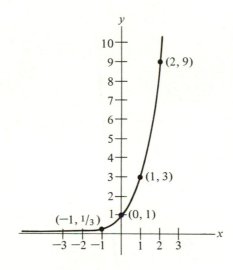

11

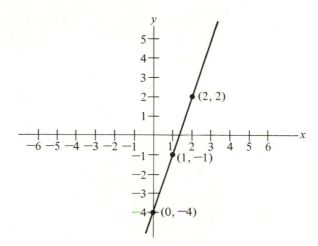

13

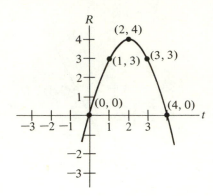

14

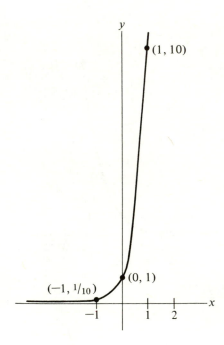

15

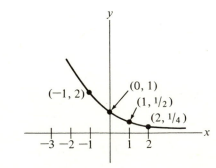

16 120

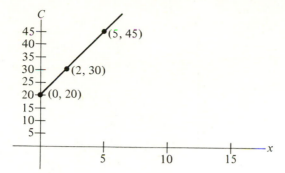

17 50 days

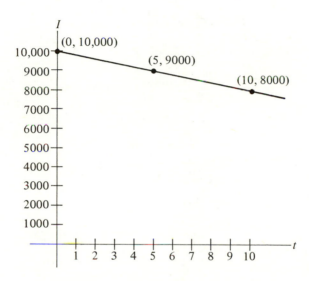

19

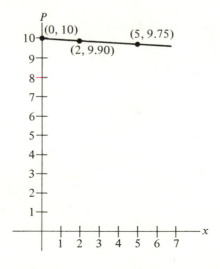

20

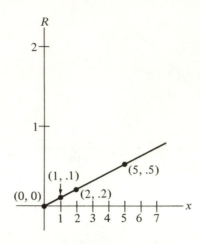

▶ **Exercise 1.2**

1 Yes   **2** Yes

4 $f(2) = 6, f(-3) = 1, f(p) = p + 4$

5 $f(0) = -1, f(3) = 8, f(-4) = 15$

7 $C(\frac{1}{2}) = 5\frac{1}{2}, C(0) = 5, C(12) = 17$

8 $R(\frac{1}{4}) = \frac{3}{4}, R(1.6) = 4.8, R(3\frac{1}{2}) = 10.5$

10 $f(2) = 4, f(0) = 3, f(-3) = 0$

11 $g(3) = -2, g(\frac{1}{2}) = \frac{1}{2}, g(0) = 3, g(-4) = -3$

13 No. The same parents may have more than one child.

14 Yes, no

16 a $f(x) = \begin{cases} .19x & \text{if } x \le 4000 \\ 760 + .22(x - 4000) & \text{if } 4000 < x \le 8000 \end{cases}$

    b $1090   **c** $380

17 a $1000   **b** $1000   **c** $f(t) = 1000$

19 205

20 $d = 50t$

    $d = 250$ miles

    $t = 2\frac{1}{2}$ hr

▶ **Exercise 1.3**

1 $x \neq 4$

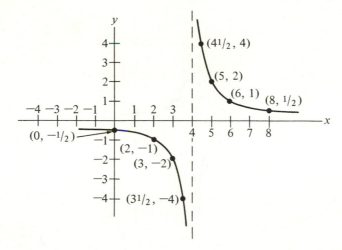

$(4^1/_2, 4)$
$(5, 2)$
$(6, 1)$
$(8, ^1/_2)$
$(0, -^1/_2)$
$(2, -1)$
$(3, -2)$
$(3^1/_2, -4)$

2 $x \geq 5$

$(14, 3)$
$(9, 2)$
$(6, 1)$
$(5, 0)$

4 All real numbers

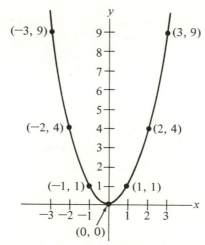

$(-3, 9)$
$(3, 9)$
$(-2, 4)$
$(2, 4)$
$(-1, 1)$
$(1, 1)$
$(0, 0)$

5 $x \neq -3$

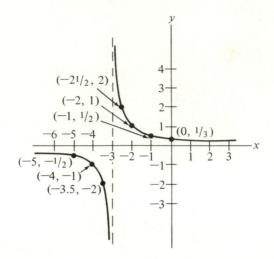

7 0, 1, 2, 3, 4, 5, . . . , 49, 50. The graph is a succession of dots with $x =$ 1, 2, . . . , 50. Some representative points are shown here.

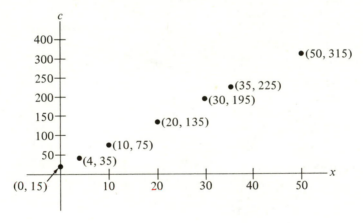

8 a 2.70   **b** 1.95   **c** 3.03

The domain is $2 \leq x \leq 4$, where x is a rational number that has been rounded off to two decimal places.

10 17, -46, the range is $-63 \leq y \leq 17$.

11 The domain is 0, 1, 2, 3, 4, . . . , 99, 100.

The range is $-100, -96, -92, -88, \ldots, 296, 300$. As for Problem 7, only a few points are shown.

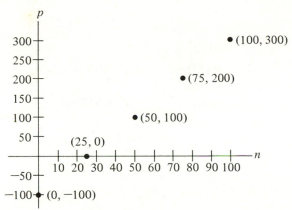

13 The range is $y \le 7$.

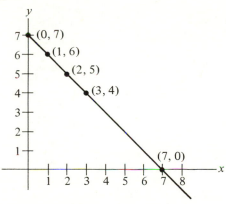

14 The range is $1, 2, 3, 4, 5, 6, \ldots$.

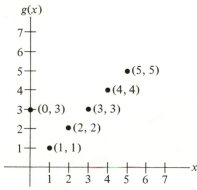

16 The graph stops at (5, 73,890).

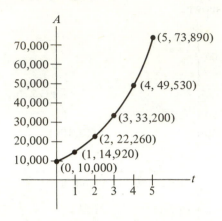

▶ **Exercise 1.4**

1 Linear; $a = -4$, $b = 1$, $c = -6$ **2** Not linear

4 Not linear   **5** Not linear

7 Not linear   **8** Not linear

10 $m = \frac{3}{2}$

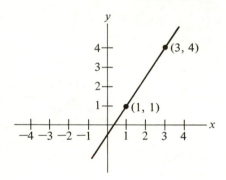

11 $m = -\frac{1}{3}$

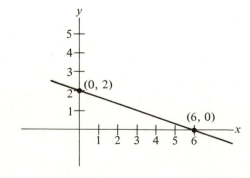

13 $m = \frac{1}{3}$

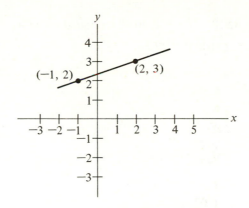

14 $m = 0$

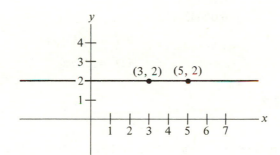

16 $m = [f(p) - f(x)]/(p - x)$

17 $m = 3$ **19** $m = 0$

20 $m = 5$, marginal cost $= 5$

22 $m = 10$ **23** $m = -4$

25 Marginal cost $= 2.13$

26 Experimental, .2 units/month; traditional, .067 units/month

▶ **Exercise 1.5**

1 $5x - y = 7$ **2** $4x + 2y = 5$

4 $4x - 3y = -2$ **5** $y = 2$

7 $5x + y = 2$ **8** $y = 1$

10 $200t - b = -7800$, $b = 9200$ bushels

11 $5v - 4w = 0$, $v = 151.2$ lb   **13** $20d + n = 1000$

14 $.2w - L = -.5$, 2.5 in., 2.5 oz

16 b $y = 1.856x + 3.828$ **c** 40.95, rounded off to 41

17 $C(u) = .75u + 75$, $C(1000) = \$825$

▶ **Exercise 2.1**

1 2.1   1.9
   2.01   1.99
   2.005   1.9995
   2.00001   1.99999

2 -5.5 -6.5
   -5.95 -6.02
   -5.998 -6.002
   -5.99875 -6.00002

7 $L = 6$, no   **8** $L = -3$, yes

10 $L = 0$, yes   **11** $L = 3$, yes

▶ **Exercise 2.2**

1 No limit

2 a Right-hand limit $= 7$ **b** Left-hand limit $= 4$ **c** No limit

4 a Right-hand limit $= 4$ **b** Left-hand limit $= 4$ **c** $L = 4$

5 Right-hand limit $= 0$

▶ **Exercise 2.3**

1 $L = 7$ **2** $L = -1$ **4** $L = 2$

5 $L = -1$ **7** $L = 1.5$ **8** $L = 0$

10 $L = \frac{15}{2}$ **11** $L = \frac{1}{6}$ **13** $L = 0$

▶ **Exercise 2.4**

1 $L = 0$ **2** $L = 1$ **4** No limit

5 $L = 0$ **7** $L = \frac{1}{2}$ **8** $L = 1$

▶ **Exercise 2.5**

1 $L = \frac{2}{3}$ **2** $L = \frac{1}{2}$ **4** No limit

5 $L = \frac{3}{2}$ **7** $L = 3$ **8** $L = 1200$

10 $L = \frac{3}{2}$ **11** $L = 0$

▶ **Exercise 2.6**

1 $2025   **2** $3639

4 a $9179   **b** $10,000   **c** 91.79%

5 $R = 3$ **7** $t = 2.5$ hr; $t = 4.016$ hr

8 $y = 40, 86.47\%$ **10** EC

11 11,050

▶ **Exercise 2.7**

1 At $x = 1, y = -1$
   $\lim_{x \to 1} (2x^2 - 3) = -1$
   $\lim_{x \to 1} (2x^2 - 3) = f(1)$, where $f(x) = 2x^2 - 3$
   Therefore the function is continuous.

2–9 These problems are worked out similarly to Problem 1.

10 $x = 3$; $g(3)$ is undefined

11 $x = 4$; y is undefined at $x = 4$

▶ **Exercise 3.1**

1 a 5.50   5
       4.38   4.6
       3.42   4.2
       3.0402   4.02 $m_{\text{sec}} \to 4$

       1.50   3
       2.28   3.6
       2.62   3.8
       2.9602   3.98 $m_{\text{sec}} \to 4$
   b $f'(x) = 4x$ **c** $f'(1) = 4$
   d Yes   **e** $m_{\text{tan}} = 4$

2 a $y' = 1$ **c** $f'(x) = 6x$ **e** $f'(x) = -1/x^2$

3 a $y = x$ **c** $12x - y = 10$ **e** $x + 4y = 4$

4 a All values of x **c** $x = \frac{1}{6}$ **e** No values of x

▶ **Exercise 3.2**

1 $f'(x) = 6x^2 - 12x + 1$ **2** $y' = -2x + 12x^2$

4 $y' = -2/x^3 - 6$ **5** $y' = 2x/5 + 6/5$

7 $y' = 1/(3\sqrt[3]{x^2})$ **8** $y' = 3\sqrt{x} - 10/x^3$

10 $f'(x) = 2x - 2/x^2$

11 $f'(x) = 20x^4 - 20x^3 + 18x^2 - 14x + 6$

13 $g'(x) = 1$ **14** $g'(x) = 26x + 4 + 1/x^2$

16 $f'(2) = 4$. This is the slope of the line tangent to $f(x) = 2x^3 - 5x^2 + 6$ at $(2, 2)$.

17 $f(0) = -7, f'(0) = 3$ **19** $(2, \frac{17}{3}), (4, \frac{13}{3})$

20 $(2, 6); 7x - y = 8$

▶ **Exercise 3.3**

1 $v = 32t$, $v = 64$ ft/sec at $t = 2$, $a = 32$ ft/sec²

2 a $dP/dn = 200 - 4n$ **b** $n = 50$
   c $0 \leq n < 50$ **d** $n > 50$

4 a $dC/dx = 2x + 8$ **c** Fixed cost $= 7$

5 $dA/dr = 2\pi r$, 10π

7 a $dV/dx = 576 - 192x + 12x^2$ **b** $x = 4$ or $x = 12$
   c $V = 1024$ cu in. at $x = 4$. It is impossible to make such a box at $x = 12$.

8 a $dP/dn = 4 - 2n$ **b** $0 \leq n < 2, n = 2, n > 2$

10 a $4x - y = -2$ **b** $4x + y = 14$ **c** $x - y = -6, x + y = 3$

▶ **Exercise 3.4**

1 Minimum point $(3, 2)$

2 Minimum point $(7, 79/3)$, maximum point $(3, 37)$

4 Minimum point $(10, 3000)$, maximum point $(50, 35,000)$

5 Minimum point $(-2, 0)$

7 Minimum point $(20, -925/3)$, maximum point $(5, 1525/6)$

8 None   **10** None

11 Minimum point $(10, -2350/3)$, maximum point $(20, -1850/3)$

13 Maximum point $(25, 625)$ **14** None

16 Absolute minimum points occur in Problems 1, 5, 9, 15.

17 Absolute maximum points occur in Problems 3, 6, 13.

▶ **Exercise 3.5**

1 Maximum revenue = 16

2 $\frac{8}{3}$ by $\frac{32}{3}$ by $\frac{32}{3}$, $V = 8192/27 = 303\frac{11}{27}$ cu in.

4 $r = 3$

5 $t = 6$, $t = 0$, $n = 114$

7 Marginal cost = $2x + 10$, average cost = $x + 10 + 4/x$, minimum average cost = 14

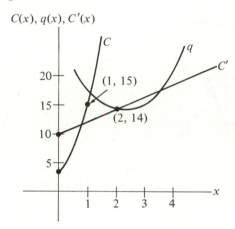

$C(x), q(x), C'(x)$

8 Marginal cost = $2x + 3$, average cost = $x + 3 + 9/x$, minimum average cost = 9

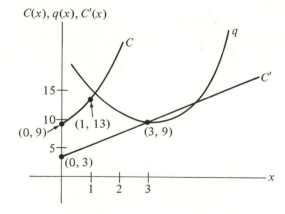

$C(x), q(x), C'(x)$

10 a $R = (450 - 1.5x)x$ **b** $x = 150$ **c** $p = 225$

11 Average revenue $= 20 + x - x^2$, marginal revenue $= 20 + 2x - 3x^2$,
$(\frac{1}{2}, 20\frac{1}{4})$

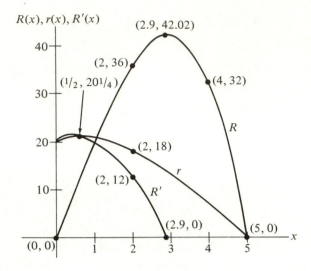

13 Average revenue $= 10 + 2x - 2x^2$, marginal revenue $= 10 + 4x - 6x^2$,
$(\frac{1}{2}, 10\frac{1}{2})$

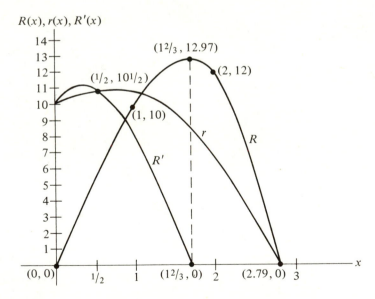

14 $d = 30$, $S(d) = 1400$

16 $Q = 2000$, Profit $= 30,000$, $p = 80$

17 $Q = 1600$, Profit $= 15,600$, $p = 84$

▶ **Exercise 4.1**

1 $y' = 20x(2x^2 + 3)^4$

2 $y' = (45x^2 - 36x)(5x^3 - 6x^2 + 7)^2$

4 $y' = (x^2 - 2)/(x^3 - 6x)^{2/3}$

5 $y' = (-48x - 28)/(6x^2 + 7x - 8)^5$

7 $C' = (10 - 10n)(2n - n^2)^4$

8 $R' = 1/\sqrt{2x + 3}$

10 $g'(x) = -6/(3x - 5)^3$

11 $f'(x) = -3/2(3x - 4)^{3/2}$

▶ **Exercise 4.2**

1 $y' = (x^2 - 3)(2) + (2x + 5)(2x) = 6x^2 + 10x - 6$

2 $y' = x^{1/2}(2x) + \frac{1}{2}x^{-1/2}(x^2 + 3) = \dfrac{5x^2 + 3}{2\sqrt{x}}$

4 $f'(x) = (3x + 2)(5) + 3(5x - 4) = 30x - 2$

5 $g'(x) = (x^2 - 1)^{1/2}(2) + 2x(\frac{1}{2}(x^2 - 1)^{-1/2}(2x)) = \dfrac{4x^2 - 2}{\sqrt{x^2 - 1}}$

7 $f'(x) = (4x + 9)(-6x^{-3} - 4x^{-2}) + 4(3x^{-2} + 4x^{-1} + 6)$

$= \dfrac{24x^3 - 48x - 54}{x^3}$

8 $y' = (x + 5)^{1/2}(\frac{1}{4})(2x - 3)^{-3/4}(2) + (2x - 3)^{1/4}(\frac{1}{2})(x + 5)^{-1/2}(1)$

$= \dfrac{3x + 2}{2(x + 5)^{1/2}(2x - 3)^{3/4}}$

10 $y' = (3x^4 - 1)^{1/2}(1/3)(2x^3 + x)^{-2/3}(6x^2 + 1)$
$\quad + (2x^3 + x)^{1/3}(\frac{1}{2})(3x^4 - 1)^{-1/2}(12x^3)$

$= \dfrac{54x^6 + 21x^4 - 6x^2 - 1}{3\sqrt{3x^4 - 1}\,\sqrt[3]{(2x^3 + x)^2}}$

▶ **Exercise 4.3**

1 $y' = \dfrac{(3x - 2)(2x) - x^2(3)}{(3x - 2)^2} = \dfrac{3x^2 - 4x}{(3x - 2)^2}$

2 $y' = \dfrac{(n + 1)(3n^2) - n^3(1)}{(n + 1)^2} = \dfrac{2n^3 + 3n^2}{(n + 1)^2}$

4 $P'(n) = \dfrac{(n^2 - 5n + 4)(2n + 6) - (n^2 + 6n - 8)(2n - 5)}{(n^2 - 5n + 4)^2}$

$= \dfrac{-11n^2 + 24n - 16}{(n^2 - 5n + 4)^2}$

5 $y' = \dfrac{(2x + 5)(2x - 6) - (x^2 - 6x + 2)(2)}{(2x + 5)^2} = \dfrac{2x^2 + 10x - 34}{(2x + 5)^2}$

7 $g'(x) = \dfrac{(x^3 + 5x)(14x + 7) - (7x^2 + 7x)(3x^2 + 5)}{(x^3 + 5x)^2}$

$= \dfrac{-7x^4 - 14x^3 + 35x^2}{(x^3 + 5x)^2} = \dfrac{-7x^2 - 14x + 35}{(x^2 + 5)^2}$

8 $y' = \dfrac{(x + 3)(\frac{1}{2})(x + 1)^{-1/2}(1) - (x + 1)^{1/2}(1)}{(x + 3)^2} = \dfrac{-x + 1}{2\sqrt{(x + 1)}(x + 3)^2}$

10 $m = \frac{5}{4}$; $5x - 4y = 3$

11 $m = 19/9$; $19x - 9y = 7$

▶ **Exercise 4.4**

1 $y' = e^x$ | **2** $y' = 2e^x$

4 $f'(x) = -\sqrt{2}(e^{-x})$ | **5** $y' = 3e^{3x}$

7 $ds/dt = 2te^{t^2}$ | **8** $q'(n) = 9e^{3n}$

10 $R'(x) = 4x^3 e^{x^4}$ | **11** $y' = (4x - 6)e^{2x^2 - 6x}$

13 $y' = e^x(x + 1)$ | **14** $y' = xe^{-x}(-x + 2)$

16 $f'(x) = 2e^{2x}(x^2 + x + 1)$ | **17** $g'(x) = e^x(x - 2)/x^3$

19 $y' = 2x + e^x$ | **20** $f'(x) = (e^x - xe^x - 1)/x^2$

22 a $(.9048)10^6$ **b** 9.5% **c** $dc/dx = -10^4 e^{-.01x}$
   d Yes   **e** $dc/dx = -9048$ at $x = 10$

23 a 18.6 ft   **b** 2.68   **c** Yes

25 a .54 miles/sec   **b** $.1e^{.5x}$ **c** .27 miles/sec/mile

26 $S = 1.84$, $s' = -.005e^{-.001x}$, yes

▶ **Summary Exercise 4.4**

1 $y' = 5(2x^4 - 6x^2 + 8)^4(8x^3 - 12x)$

2 $y' = -56x^6 - 72x^5 + 6x^2 + 6x$

4 $Q'(n) = \dfrac{(n^2 + 10n + 5)}{(n + 5)^2}$

5 $y' = (5x - 1)^2(30x^3 - 3x^2 + 40x - 2)$

7 $y' = e^{-x}(-x + 1)$

8 $y' = \dfrac{e^{-x}(-e^{x^2} - x - 2xe^{x^2} - 1)}{(e^{x^2} + x)^2}$

▶ **Exercise 4.5**

1 a $R = 10ne^{-.005n}$ **b** $n = 200$ **c** $S = 3.68$

2 a $dv/dt = -10,000 + 200t$ **b** 250,000   **c** 50   **d** 50 hr

4 $R = 2xe^{-x/100}$; $x = 100$

5 $l = 20$; $w = 30$; 120 yd

7 $l = w = h = 6$, $A = 216$

8 a $C = 7L/2 + 35,000/L$ **b** $L = 100$, $w = 350/3$ **c** \$700

10 4 by 4 by 4

11 $x = 225$

13 $V = (15 - 2x)(8 - 2x)(x)$, 5/3 by 14/3 by 35/3

▶ **Exercise 4.6**

1 $y = 0$, $x = 0$ **2** $y = \frac{3}{2}$, $x = -1$ and $x = \frac{5}{2}$

4 None, $x = \frac{1}{2}$ **5** $y = 5$, $x = 3$ and $x = -3$

7

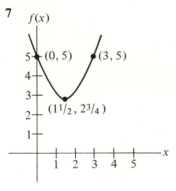

8

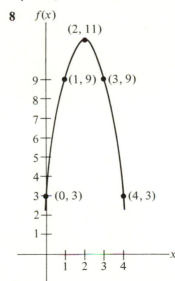

10

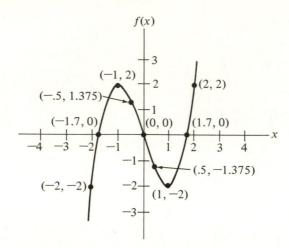

11

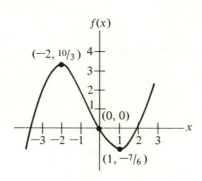

13

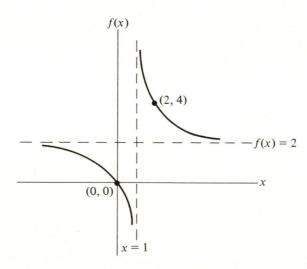

14

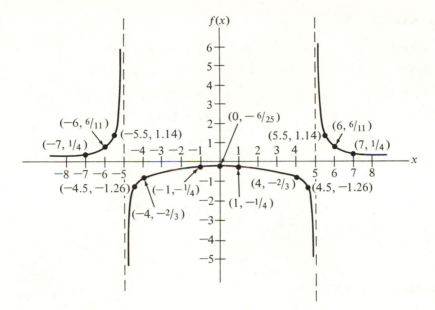

$(-6, {}^6/_{11})$
$(-5.5, 1.14)$
$(-7, {}^1/_4)$
$(0, -{}^6/_{25})$
$(5.5, 1.14)$
$(6, {}^6/_{11})$
$(7, {}^1/_4)$
$-8\ -7\ -6\ -5$
$-4\ -3\ -2\ -1$
$1\ \ 2\ \ 3\ \ 4$
$5\ \ 6\ \ 7\ \ 8$
$(-4.5, -1.26)$
$(-1, -{}^1/_4)$
$(4, -{}^2/_3)$
$(4.5, -1.26)$
$(-4, -{}^2/_3)$
$(1, -{}^1/_4)$

16

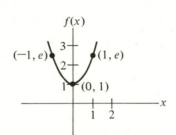

$(-1, e)$
$(1, e)$
$(0, 1)$

17

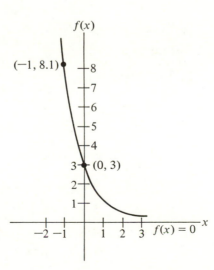

$(-1, 8.1)$
$(0, 3)$
$f(x) = 0$

▶ Exercise 5.2

1 $2x + c$

2 $-5w + c$

4 $2x^3/3 + 3x^2/2 - 5x + c$

5 $3x - x^2 + x^5/5 + c$

7 $2x^{3/2}/3 + c$

8 $4x^{3/2}/3 + c$

10 $y^3/3 - 3y^2 + c$

11 $-1/w + \frac{3}{4}w^{4/3} + c$

13 $x^3/3 + bx^2/2 + c$

14 $(5x^3 - 5)^5/75 + c$

16 $(x^2 + 3x^3)^5/5 + c$

17 $x^4/4 + 3x^2/2 + c$

19 $(2x^3 + 1)^{100}/600 + c$

20 $(5x^3 + 1)^7/105 + c$

22 $2(3x + 4)^{3/2}/9 + c$

23 $(9x^2 + 2)^{4/3}/24 + c$

25 $(x + 2x^2)^2/2 + c$

▶ Exercise 5.3

1 $e^x + c$

2 $e^{3x} + c$

4 $e^{3x^4} + c$

5 $e^{x^2-2x} + c$

7 $2e^x + c$

8 $e^{5x^2}/10 + c$

10 $-e^{-x} + c$

11 $-50e^{-.2t} + c$

▶ Exercise 5.4

1 $2e^x(x - 1) + c$

2 $e^x(x - 1)/5 + c$

4 $x(x - 1)^6/3 - (x - 1)^7/21 + c$

5 $2x(x + 2)^{1/2} - 4(x + 2)^{3/2}/3 + c$

7 $3x(2x + 7)^{4/3}/8 - 9(2x + 7)^{7/3}/112 + c$

8 $e^x(x^2 - 2x + 2) + c$

▶ Exercise 5.5

1 $x^4/4 + e^{3x} + c$

2 $(x^3 + e^{2x})^3/3 + c$

4 $e^x + 2x + c$

5 $e^x + ex + c$

7 $2(3x + 4)^{1/2}/3 + c$

8 $-3e^x + c$

10 $-e^{-x} + c$

11 $(x^2 - 5)^{1/2} + c$

▶ **Exercise 5.6**

1 $C(x) = 25x^2 + 1000x + 2000$; $C(50) = 114{,}500$

2 $S = 320t - 16t^2$, velocity zero at $t = 10$, maximum height $= 1600$

4 family $f(x) = x^2 + 3x + c$; member $f(x) = x^2 + 3x - 4$

5 $S = -250e^{-.02t} + 750$

7 $f(x) = 2x^3/3 - 3x^2 + 25/3$

8 $Q(x) = x^2 + 3x$, 2650

10 $P = 200x - 5x^2 - 500$, $P = 1500$

11 a $R(x) = 150x$
  b $C(x) = x^2 + 17x + 1000$
  c (8, 1200), (125, 18,750)
  d $P(x) = 133x - x^2 - 1000$
  e -1000
  f $66\frac{1}{2}$
  g 8 through 125

13 $R = .08t^{5/4} + 1$, $R = 3.56$

14 $v = 50e^{-t}$, $v = 2.5$

16 $h = 10t - 920$, $h = 80$

▶ **Exercise 5.7**

1 $R = .3P + c$

2 $y = 3x^2/2 - 4x + c$

4 $y^3/3 = x^2/2 + c$

5 $y = x^4 + c$

7 $-1/y^2 = x^2 + c$

8 $v = 20t$, $v = 20$

▶ **Exercise 6.1**

1 2/3

2 6.389

4 $-44/3$

5 4/5

7 189/12

8 $(e - 1)/2$

▶ **Exercise 6.2**

1 10

2 9

4 12

5 9/2

7 15/4

8 56/3

10 18

11 78

13 2.218

14 1.718

16 32/3

17 32/3

19 7/2	20 3/8	22 6
23 3/2	25 6	26 1
28 1	29 5	

▶ **Exercise 6.3**

1 1/6	2 $4\frac{1}{2}$	4 $17 - e^2$
5 16/3		

▶ **Exercise 6.4**

1 $\displaystyle\int_{2}^{3} 1 \, dx = 1$ 2 $k = 2/9$

4 $k = 3/7; 12.5\%$ 5 $4343\frac{3}{4}$

7 4/3   8 2/3

▶ **Exercise 6.5**

1 Converges to $\frac{1}{2}$ 2 Diverges

4 Converges to 4   5 Diverges

7 a $28,571.43   b $25,000

▶ **Exercise 6.6**

1 20.5   2 .643

4 5.34375   5 6.697

7 $A = 20$; error $= .5$ 8 $A = .666\ldots$; error $= .0236\ldots$

▶ **Exercise 7.1**

1 $y' = 30x^4 + 12x^2, y'' = 120x^3 + 24x$

2 $f'(x) = 3x^2 + 8x - 6, f''(x) = 6x + 8$

4 $y' = -1/x^2, y'' = 2/x^3$

5 $y' = \frac{1}{2}(x + 1)^{-1/2}, y'' - \frac{1}{4}(x + 1)^{-3/2}$

7 $y' = e^x, y'' = e^x$

8 $y' = -2e^{-2x}, y'' = 4e^{-2x}$

10 $(-3, -26)$ is a minimum point.

11 $(-3, 19)$ is a maximum point.

13 $(-1/6, 25/12)$ is a maximum point.

14 $(0, 0)$ is a maximum point, $(6, -108)$ is a minimum point.

16 $(0, -1)$ is a maximum point.

17 Minimum average cost $= 200$.

19 $x = 32, y = 32$

20 $(0, .135)$ is a minimum point.

22 2000 units

▶ **Exercise 7.2**

1 $C_x = 4x + n, C_{xx} = 4, C_n = -9n^2 + x, C_{nn} = -18n, C_{xn} = C_{nx} = 1$

2 $Z_x = 4xy^5 - 3x^2y, Z_{xx} = 4y^5 - 6xy, Z_y = 10x^2y^4 - x^3,$
   $Z_{yy} = 40x^2y^3, Z_{xy} = Z_{yx} = 20xy^4 - 3x^2$

4 $Z_x = 2y + 6xy + y^3, Z_{xx} = 6y, Z_y = 2x + 3x^2 + 3xy^2, Z_{yy} = 6xy,$
   $Z_{xy} = Z_{yx} = 6x + 2 + 3y^2$

5 $Z_x = 1/y^2 - y, Z_{xx} = 0, Z_y = -2xy^{-3} - x, Z_{yy} = 6xy^{-4},$
   $Z_{xy} = Z_{yx} = -2y^{-3} - 1$

7 $Z_x = 2e^{2x} + y, Z_{xx} = 4e^{2x}, Z_y = x, Z_{yy} = 0, Z_{xy} = Z_{yx} = 1$

8 $Z_p = e^p + 4pq^3, Z_{pp} = e^p + 4q^3, Z_q = 6p^2q^2, Z_{qq} = 12p^2q,$
   $Z_{pq} = Z_{qp} = 12pq^2$

10 $z_l = 6(l)^2 - m^2 + 3m, z_m = -2(lm) + 3(l)$

11 $D_n = 2n - 3p, D_p = -3n + 2p$

13 $f_{m_1} = gm_2/d^2, f_{m_2} = gm_1/d^2, f_d = -2gm_1m_2/d^3$

14 $I_R = -V/R^2, I_V = 1/R$

▶ **Exercise 7.3**

1 $(-2, 5), z = -29$ is a minimum value.

2 $(\frac{1}{2}, 1), z = -\frac{1}{4}$ is a minimum value.

4 $(4, 0), z = -29$ is a minimum value.

5 At $(0, 0)$ the test fails.

7 $(72, 12), z = 1000$ is a minimum value.

▶ **Exercise 7.4**

1 $-2x/3y$

2 $-(x + 3y)/(3x + y)$

4 $(1/x^2 - y + 2)/(x - 3y^2)$

5 $m = -3; y = -3x + 4$

▶ **Exercise 7.5**

1 $(15, 15)$ is a critical point, $z = 1125$ is a maximum value.

2 $(10, 10)$ is a critical point, $z = 100$ is a minimum value.

4 $(1, 1, -1)$ is a critical point, $q = 3$ is a minimum value.

5 $(9/91, 54/91, 108/91)$ is a critical point, $q = 14{,}642/8281$ is a minimum value.

7 $(-20, 20)$ is a critical point, $z = 0$ is a minimum value.

8 $(5, 15)$ is a critical point, $C = 350$ is a minimum value.

▶ **Exercise 8.1**

1

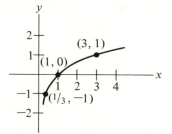

2

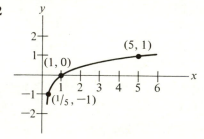

4

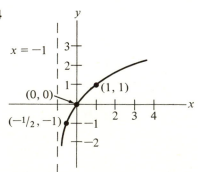

5

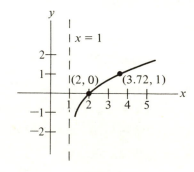

▶ **Exercise 8.2**

1 $1/x$

2 $(2x + 3)/(x^2 + 3x - 4)$

4 $3/(x + 1)$

5 $14/(2x + 5)$

7 $-7/(x - 3)(2x + 1)$

8 $1/(x + 1) + 2/(2x + 7)$

10 Approximately 39.6 ft; $.1y(105 - y)/105$

11 No; $dy/dx \neq 0$

13 $1/x \log_e 10$

14 $3/(3x - 1) \log_e 10$

16 $(2x - 3)/(x^2 - 3x + 4) \log_e 5$

17 $3x^2/(x^3 - 6) \log_e 10$

19 $1/(x + 3) \log_e 4$

20 $3/\log_e 3$

▶ **Exercise 8.3**

1 $x^{2x}(2 + 2 \log_e x)$

2 $(x + 1)^x [x/(x + 1) + \log_e (x + 1)]$

4 $(x + 1)^{2x} [2x/(x + 1) + 2 \log_e (x + 1)]$

5 $(e^x)^{e^x}(e^x + xe^x)$

7 $x^{3x}(3 + 3 \log_e x)$

8 $3^x(\log_e 3)$

▶ **Exercise 8.4**

1 $\log_e (x^2 + 5) + c$

2 $\log_e (x^3 - 10) + c$

4 $\log_e (e^x - 1) + c$

5 $\log_e (2x + 3) + c$

7 $\frac{1}{6} \log_e (2t^3 + 3) + c$

8 $\log_e (x + 1) + c$

▶ Exercise 8.5

1 $\log_e x = 2t + c$

2 $\log_e v = .2t + c$

4 $dI/dt = kI$, $\log_e I = kt + c$

5 $dA/dt = kA$, $\log_e A = kt + c$

7 135

8 $c \doteq 3.9$, $k = \log_e 2$

▶ Exercise 8.6

1 $P = 50 \log_e (x + 1) - 100$; $P \doteq 200$

2 $\frac{1}{2} \log_e 2$

4 $\log_e 5 - \log_e 3$, or $\log_e (\frac{5}{3})$

5 Diverges

7 Diverges

13 $1/(3 - x)$

14 $(-2x^2 + x - 3)/(x^2 + 1)(x + 1)$

16 $(\frac{1}{2}, e^{-2})$

17 a $z_x = 2/(x + y)$, $z_y = 2/(x + y)$

   b $z_x = 2x^3/(x^2 + y) + 2x \log_e (x^2 + y)$, $z_y = x^2/(x^2 + y)$

▶ Exercise 1, Appendix 1

1 x^{10}	**2** y^{a+1}	**4** x^{12}
5 y^{2a-1}	**7** $(x + 1)^{y+a+3}$	**8** 10^4

▶ Exercise 2, Appendix 1

1 1	**2** $1/x^3$	**4** $4b^2/a^3$
5 x^{10}	**7** x^{-1}	**8** $3x^{-2}$
10 $x^{-1}y^{-1}$	**11** $1/x^{-3}$	

▶ Exercise 3, Appendix 1

1 $\sqrt{25} = 5$	**2** $1/\sqrt{25} = 1/5$	**4** $1/27$
5 4	**7** $x^{1/3}$	**8** $(x^2 + 4)^{1/2}$
10 $3(x + 5)^{-1/3}$	**11** $x(x^2 - 4)^{-1/2}$	

▶ **Exercise 1, Appendix 2**

1 Exponent, logarithm, 4

2 Exponent, logarithm, base

4 Exponent, logarithm 4, base, 8

5 Exponent, e, logarithm, 1.105, base, e

7 Ten, exponent, 100, 10, 100

8 Yes, $a^0 = 1$ for all positive a

10 $\log_2 16 = 4$ **11** $\log_4 64 = 3$ **13** $\log_{10} 1000 = 3$

14 $\log_{10} .1 = -1$ **16** $\log_e .1353 = -2$ **17** $\log_{16} \frac{1}{2} = -\frac{1}{4}$

19 $2^5 = 32$ **20** $25^{1/2} = 5$ **22** $e^{2.5} = 12.182$

23 $9^{-2} = 1/81$

▶ **Exercise 2, Appendix 2**

1 3   **2** 7.389   **4** 1/5

5 3   **7** 4   **8** 3

10 $x = 11$ **11** $x = 4$

▶ **Exercise 3, Appendix 2**

1 9   **2** 4   **4** 7/3

5 $\log_a x + \log_a y$ **7** $\log_a (xy)$ **8** $\log_2 27$

10 $\frac{1}{2}$ **11** 2   **13** -3

14 $\log_a x - \log_a y$ **16** $\log_a x/y$ **17** $\log_5 27/10$

19 4   **20** 6   **22** $\log_5 5^2 = 2$

23 $\log_b x^a$ **25** $x = 2$ **26** $x = 2$

▶ **Exercise 4, Appendix 2**

1 $\log_a x + \log_a y$ **2** $2 \log_a x + 3 \log_a y - \log_a z$

4 $\log_a (x^3/y)$ **5** $\log_a (xy^2/z^3)$

7 $x = -1$ **8** $x = 14$

▶ **Exercise 5, Appendix 2**

1 $\frac{3}{4}$ **2** $\frac{3}{2}$ **4** 2

5 $y = \dfrac{\log_e x^3}{\log_e 10}$ **7** $y = \dfrac{\log_e (2x + 4)}{\log_e 10}$ **8** $y = \dfrac{\log_e (x + 3)^5}{\log_e 10}$

Chapter One

1 $f(0) = 3, f(3) = 12, f(-2) = 7$

2 $f(3) = 5, f(6) = 15, f(1) = 1$

3 $P(400) = 600; n = 160$
  The domain is 0, 1, 2, 3, 4, . . . , 1000.
  The range is $-400, -397.50, -395, -392.50, . . . , 2100.$

4 $\frac{10}{3}$

5 $3x - 2y = 0$

6

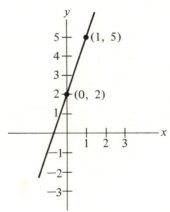

7 a $x \leq 100$ **b** $x \neq -\frac{4}{3}$

Chapter Two

1 $L = 5$ **2** $L = -2$ **3** $L = \frac{2}{3}$

4 $L = 0$ **5** $L = 140$ **6** No limit

7 a Right-hand limit $= 13$ **b** Left-hand limit $= -5$ **c** No limit

8 $x = 5$

9 Yes; $f(1) = 1$
$$\lim_{x \to 1} f(x) = 1$$
$$f(1) = \lim_{x \to 1} f(x)$$

Chapter Three

1 $f'(x) = 2x$ **2** $f'(x) = 3 - 2x$

3 $f'(x) = -60/x^3$ **4** $y' = 2/\sqrt{x}$

5 $f'(x) = 12x^2 - 6x$ **6** $(2, 6)$

7 Fixed cost $= 700$; marginal cost function is $C'(x) = 6x + 4$; marginal cost $= 28$

8 a $v = 224$ ft/sec   **b** 448 ft   **c** 1024 ft

9 1200 units; maximum revenue $= 360{,}000$

Chapter Four

1 $dy/dx = (2x^4 + 6x)^{-1/2}(4x^3 + 3)$

2 $dy/dx = [(3x - 2)(2x) - (x^2 + 3)(3)]/(3x - 2)^2$ (unsimplified)
$dy/dx = (3x^2 - 4x - 9)/(3x - 2)^2$ (simplified)

3 $dy/dx = -3e^{-3x+1}$

4 $dy/dx = 2x^2 e^{x^2} + e^{x^2}$ or $dy/dx = e^{x^2}(2x^2 + 1)$

5 a $x = 2$ **b** $y = 0$ **c** None

d

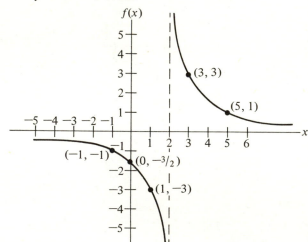

6 a None   **b** None   **c** $(3, -8)$ is a minimum; $(\frac{1}{3}, \frac{40}{27})$ is a maximum

d

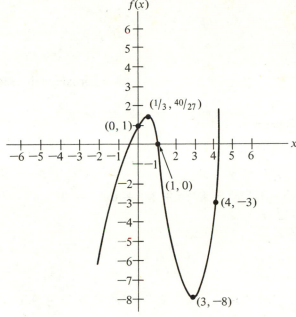

7 $x = 25$ **8 a** $R(x) = 50x - .02x^2$ **b** $x = 1250$

Chapter Five

1 $\frac{2}{3}x^{3/2} + c$

2 $x^3 + 5x/2 - 6x + c$

3 $-3/x + c$

4 $\frac{1}{3}(x^2 + 3)^{3/2} + c$

5 $\frac{4}{3}(3x - 4)^{1/2} + c$

6 $\frac{1}{6}e^{3x^2+1} + c$

7 $2x - 3e^{-x} + c$

8 $-xe^{-x} + e^{-x} + c$

9 $C(x) = \frac{3}{2}x^2 + 5x + 1000$

10 $y = \frac{1}{2}x^2 - 5x + c$

11 $\frac{1}{4}y^4 = x^2 + c$

12 $S = -250e^{-.02t} + 350$

Chapter Six

1 $\frac{23}{6}$

2 $\frac{1}{2}(e^4 - 1)$

3 2

4 $\frac{11}{3}$

5 $20\frac{5}{6}$

6 3

7 9

8 Approximately .364

Chapter Seven

1 $y' = 2x + 6$, $y'' = 2$

2 $y' = 1 - 1/x^2$, $y'' = 2/x^3$

3 $y' = 2xe^{x^2} + 3x^2$, $y'' = 4x^2e^{x^2} + 2e^{x^2} + 6x$

4 $y' = \dfrac{-1 - y^3}{3xy^2 + 2y}$

5 $y' = \dfrac{-e^y - ye^x}{e^x + xe^y}$

6 $Z_x = 2x + 3y^3$, $Z_{xx} = 2$, $Z_{xy} = Z_{yx} = 9y^2$, $Z_y = 2y + 9xy^2$, $Z_{yy} = 2 + 18xy$

7 A minimum occurs at $(0, -1)$, minimum value $= -6$

8 A maximum occurs at $(12, 14)$, maximum value $= 74$

9 A maximum occurs at $(2, 4)$, maximum value $= -48$

Chapter Eight

1

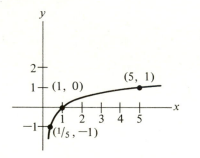

2

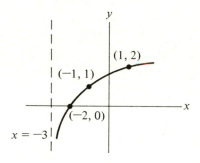

3 $y' = 1/x$

4 $y' = 2x$

5 $dy/dx = 1/x(x + 1)$

6 $dy/dx = 1/3(x + 2)$

7 $y' = 1/x \log_e 2$

8 $dy/dx = 2x \log_{10} e$

9 $y' = 5x^{5x}(1 + \log x)$

10 $y' = \log_e 4(4^{x-1})$

11 $3 \log_e (x + 1) + c$

12 $\frac{3}{4} \log_e (2x^2 + 5) + c$

13 $\frac{1}{2} \log_e (2x + 1) + c$

14 $\frac{1}{2} \log_e Q = t + c$, or $Q = e^{2(t+c)}$

15 $\log_e y = x^2/2 + c$, or $y = e^{(1/2)x^2+c}$

16 $\log_e h = .05t + 3$, or $h = e^{.05t+3}$

17 $C = \$3600$

Appendix 1

1 $x; 4$

2 $y; 2$

3 $6y; 2$

4 x^{6+b}

5 x^4

6 x^{2a-6}

7 $1/x^3$

8 $3/x^4$

9 x^7

10 $x^{-1/2}$

11 $1/x^{-5}$

12 $(x - 1)^{-1}$

13 $\sqrt[3]{x^5}$

14 $1/\sqrt[4]{a}$

15 $3\sqrt{x - 1}$

16 $x^{1/5}$

17 $(x^2 + 25)^{1/2}$

18 $2/(c + 7)^{1/3}$

Appendix 2

1 Logarithm, 8, exponent

2 $\log_5 25 = 2$

3 $9^{-2} = 1/81$

4 a $x = \frac{1}{8}$ **b** $x = 9$ **c** $x = -1$ **d** $x = -2$ **e** $x = 2$

5 $\log_b 3 + 2 \log_b x - \log_b y$

6 $\log_7 50 - \log_7 73$

7 $\log_b (xy/z)$

8 a $x = 4$ **b** $x = 6$

9 3.9

10 $y = (\log_e x / \log_e 10)$

index